Building with
Reclaimed Components and Materials

Building with Reclaimed Components and Materials

A Design Handbook for Reuse and Recycling

Bill Addis

London • Sterling, VA

First published by Earthscan in the UK and USA in 2006
Reprinted 2007

ISBN-13: 978-1-84407-274-3

Typesetting by FISH Books
Printed and bound in the UK by MPG Books Limited, Bodmin
Cover design by Yvonne Booth

Cover images: top left, photo by BSRIA; top right, Duchi shoe shop, The Netherlands (by courtesy of 2012 Architecten); bottom left, Canalside West, Huddersfield, UK (by courtesy of Ove Arup); bottom right, photo by Buro Happold.

For a full list of publications please contact:

Earthscan
8–12 Camden High Street
London, NW1 0JH, UK
Tel: +44 (0)20 7387 8558
Fax: +44 (0)20 7387 8998
Email: earthinfo@earthscan.co.uk
Web: **www.earthscan.co.uk**

22883 Quicksilver Drive, Sterling, VA 20166-2012, USA

Earthscan is an imprint of James and James (Science Publishers) Ltd and publishes in association with the International Institute for Environment and Development

A catalogue record for this book is available from the British Library

Library of Congress Cataloging-in-Publication Data

Addis, William, 1949–
Building with reclaimed components and materials: a design handbook for reuse and recycling/Bill Addis.
p. cm.
Includes bibliographical reference and index.
ISBN-13: 978-1-84407-274-3
ISBN-10: 1-84407-274-6
1. Building materials. 2. Recycling (Waste, etc.) I. Title.
TA403.6.A37 2006
691–dc22

200503263

Printed on recycled paper with a minimum of 75 per cent de-inked post-consumer waste content.

Contents

Figures, Tables and Boxes

Figures

Tables

Boxes

Foreword

by Thornton Kay

In recent years, the proportion of construction materials being reclaimed and reused has been falling. The proportion of reusable materials being recycled or sent to landfill is now around 50 per cent higher than it was ten years ago.

As an example, some 700,000 tonnes of reusable wood were reclaimed in Britain in 1998 – around 10 per cent of the total annual demand. Since then the amount of timber being reclaimed has dropped dramatically and is still falling. Meanwhile hundreds of millions of pounds are being poured into the recycling sector. The consequence is that much irreplaceable, first-growth forest timber is now being chipped to produce tax-subsidized mulch, seen in sad bags in garden centres and supermarkets, at a very much higher fossil fuel energy cost than the human energy needed for reclaiming. This contributes to greatly increased carbon emissions. Another consequence is that the reclaimed timber trade in UK now has to look abroad to Europe, the US and even further afield to fulfil the strong private-sector demand for supplies of reclaimed timber. A similar picture exists for other countries where recycling of timber is being promoted in preference to reclamation, and for reclaimed materials sectors too.

During recent years construction professionals have had their own problems, struggling to keep pace with advances in building and technology. This has led to the increasingly complex topic of reclaimed materials being left at the bottom of a growing pile of ever more time-consuming choices.

But now, at last, reuse has begun to figure on the construction radar. In response, the salvage and reclamation trades must evolve from their cosy love affair between maverick dealers and upmarket homeowners, to one whose strong environmental credentials can harness the growing interest of government and the mainstream construction industry.

Undoubtedly difficulties exist. The vagaries of the reclaimed materials supply chain, question marks over standards of goods supplied, the enthusiasm, or not, of the demolition and dismantling sector, issues over performance, indemnity insurance, standards and regulations – to name a few – stand in the way of a smooth path to greater reuse by professionals.

My mantra has been that preventing old materials from being recycled or sent to landfill, and reusing reclaimed materials in buildings and landscapes, are both client-led. If the client is keen, professionals and their contractors will make it happen. I believe it will still be a long time before professionals will feel not only comfortable about reuse, but also keen to promote it to clients. A move in that direction must start somewhere. The acceptance of materials reuse in mainstream professional construction is in its infancy, and needs exposure and the reassurance of its peers.

This book by Bill Addis is just the type of thing that will help the construction professions develop greater confidence in the idea of reusing old materials, and will, I hope, result in greater quantities of valuable material resources being rescued from destruction, more energy being saved and carbon emissions reduced. I hope too that the book will contribute to greater interest in this important topic, which may in turn spark the reuse revolution that will inevitably happen.

Thornton Kay
Senior Partner, Salvo Llp, UK

Acknowledgements

The research leading to the writing of this book was funded by the Department of Trade and Industry as a Partners in Innovation Award.

Authors

The work was carried out by a consortium led by Buro Happold and including the Building Services Research and Information Association (BSRIA), James and James (Science Publishers) Ltd and National Green Specification (NGS). The book was compiled and written by Bill Addis (Buro Happold) with a major contribution on building services from Roderic Bunn (BSRIA). Many valuable comments and suggestions were made by Brian Murphy (NGS). Additional research and comment were provided by staff of Buro Happold and BSRIA.
The work was undertaken with guidance from the following steering group and the research consortium is grateful to members for their advice on the study and comments on the draft reports.

Appendix A is included by kind permission of CIRIA. It is taken from *Design for Deconstruction: Principles of Design to Facilitate Reuse and Recycling*, Publication C607 (CIRIA, 2004).

Steering group members

Mr Roderic Bunn	BSRIA
Mr Guy Robinson	James & James (Science Publishers) Ltd
Mr Brian Murphy	NGS GreenSpec
Mr Graham Vincent	Bovis LendLease
Mr Andrew Kinsey	Bovis LendLease
Mr Hans Haenlein	Hans Haenlein Architects
Ms Anne Chick	Kingston University
Mr Martin Long	Stanhope plc
Ms Anna Scothern	The Concrete Centre
Mr Mike Duggan	Federation of Environmental Trade Associations
Mr Graham Raven	The Steel Construction Institute
Dr Hugh St John	Geotechnics Consulting Group
Mr Martin Hunt	Construction Industry Environmental Forum

List of Acronyms and Abbreviations

BBA	British Board of Agrément
BedZED	Beddington Zero Energy Development
BMS	building management systems
BOS	basic oxygen steel-making process
BRE	Building Research Establishment
BREEAM	Building Research Establishment Environmental Assessment Method
BSRIA	Building Services Research and Information Association
CCTV	closed-circuit television
CDM	construction, design and management
CFC	chlorofluorocarbon
CHP	combined heat and power
COSHH	Control of Substances Hazardous to Health
DfD	design for deconstruction/disassembly
DIY	do-it-yourself
DPC	damp-proof course
DX	direct expansion
EAF	electric arc furnace
EPDM	ethylene propylene diene monomer
ELVs	end-of-life vehicles
ERC	Energy Resource Centre
FSC	Forestry Stewardship Council
FTSE	Financial Times Stock Exchange
GBC	Green Building Challenge
GGBS	ground granulated blastfurnace slag
GJ/t	GigaJoules/tonne
GRC	glass-reinforced cement
GRP	glass-reinforced polymer
HCFC	hydrochlorofluorocarbon
HDPE	high-density polyethylene
HSFG	high-strength friction-grip
HV	high voltage
HVAC	heating, ventilation and air-conditioning
IISBE	International Initiative for a Sustainable Built Environment
LCA	life-cycle analysis
LDPE	low-density polyethylene
LEED	Leadership in Energy and Environmental Design
MDF	medium-density fibreboard
MDPE	medium-density polyethylene
MRF	materials recycling/recovery facility
MWSF	mixed waste sorting facilities
NGS	National Green Specification

NHBC	National House-Building Council
O&M	operation and maintenance
PBB	polybrominated biphenyl
PBDE	polybrominated diphenyl ether
PET	polyethylene terephthalate
PFA	pulverized fuel ash
PP	polypropylene
PS	polystyrene
PVC	polyvinyl chloride
RCA	recycled concrete aggregate
RCBP	recycled-content building product
ROHS	Restriction of Hazardous Substances (2003)
SPG	Supplementary Planning Guide
TRADA	Timber Research and Development Association
UPS	uninterruptible power supplies
WEEE	Waste Electrical and Electronic Equipment (directive, 2003)

Introduction

The need for the design handbook

It is increasingly common for building designers and their clients to express an intention, or at least a wish, to use recycled materials or various 'second-hand' components or equipment in construction work. This has long been undertaken in small-scale 'domestic' building work and in some 'heritage' projects, especially by making use of materials and goods available through the architectural salvage market. However, little progress has been made on incorporating reclaimed goods and materials into large projects undertaken by mainstream architects, design engineers and contractors.

Recently, a growing number of policies and initiatives at local, regional and national government levels in many countries are aimed at encouraging the use of reclaimed products and materials on a larger scale.

The waste industry has generated much information and guidance with the aim of reducing the quantities of waste being sent to landfill, generally by promoting uses of 'waste' materials such as crushed concrete and bricks, crushed glass bottles and used vehicle tyres. The recycling industry has produced useful information aimed primarily at stimulating the markets for products including some construction products made from recycled materials such as plastics and rubber. Such information and guidance, however, does not directly address the needs of building designers, whether architects or engineers, by helping them incorporate reclaimed materials and components in building construction in the buildings they design. This book aims to fill this gap.

Designing buildings to incorporate reclaimed products and materials

In one respect, designing buildings to incorporate reclaimed products and materials is fundamentally different from conventional design methods. In conventional design, the designer conceives the elements and systems of a building and then specifies the materials and components needed to achieve the desired building performance and quality. Generally there already exists an established market for suitable materials and components and they can be easily purchased.

When designing buildings to incorporate reclaimed products and materials, no equivalent established market exists. It becomes virtually essential for the project team to identify the source of suitable materials and products before detailed design can commence, and before specification and tendering is undertaken.

The book considers three main types of reuse and recycling: the reuse, in situ, of a whole building or some of its parts; the reuse of components that have been removed from one building, then refurbished or reconditioned and purchased for use in a different building; finally there is the use of recycled materials, for example in what are known as recycled-content building products (RCBPs).

Medium to large building projects present a range of challenges, especially due to the quantities of goods and materials involved and their availability, the need for warranties, the complexity of the engineering involved, and the size and make-up of the project team.

During the last decade or so demolition methods have moved away from careful dismantling towards more brutal processes that tend to reduce buildings to piles of rubble. It would clearly make more (environmental) sense to salvage all components before they have been destroyed, so they might be reconditioned and reused. This book does not, however, cover the theme of 'design for deconstruction' that addresses how buildings should be designed and constructed in order that they can be easily deconstructed to facilitate reuse and recycling; this subject has been dealt with elsewhere (Addis and Schouten, 2004).

The Nomadic Museum, New York City, 2005 – Architects: Shigeru Ban and Dean Maltz; engineers: Buro Happold

Note: The walls of this travelling exhibition building are made of 148 rented shipping containers. The roof is supported by columns made of cardboard tubes 760mm in diameter, made from recycled paper. The interior structure and fittings pack into 37 containers for transporting to a new location.

Source: The photo is used with permission from the photographer, Michael Moran

This handbook addresses the key issues related to using reclaimed equipment, components and materials in building projects. It identifies the main issues that need to be considered when designing and procuring a building that will incorporate a significant proportion of reclaimed and recycled goods and materials. It directs the reader to the many players in the world of reclamation and recycling who may be unfamiliar to designers and contractors who usually work with new products and materials – for example, demolition and reclamation contractors, waste contractors, the architectural salvage sector, reconditioning businesses and the growing number of industries that make products using recycled materials.

Target readership

The book is intended for all those who wish to increase the use of reclaimed materials and products in new construction work:

- building clients and developers;
- designers, including architects, engineers, interior designers and landscape architects;
- project, design and environmental managers for building projects;
- contractors and specialist subcontractors involved in both construction and demolition work;

- planners and those implementing environmental policy;
- building and quantity surveyors;
- specification providers, contract lawyers and Professional Indemnity insurers;
- regulators and building inspectors;
- businesses dealing with reclaimed goods and materials recycling;
- manufacturers of components and equipment for the building industry.

The book does not aim to persuade those sceptics who do not yet believe that something must be done to reduce the environmental impact of our industrial society and to reduce the resources it consumes and discards. Rather, it assumes that we need to vigorously pursue the goal of a 'zero waste' society as we are now starting on the long process of reducing our use of fossil fuels in order to achieve a 'zero carbon economy' that causes no net increase in carbon dioxide in our atmosphere.

How to use the handbook

The first three chapters of the book provide a general overview of reclamation and recycling. The first chapter is an introduction to the vocabulary and concepts in common use and to the main players in the commercial market place of waste, reclamation and recycling. Chapter 2 provides some case studies that indicate something of the range of what is already being done. The third chapter provides guidance on how to ensure that reclamation, reuse and recycling are incorporated into a building project. Together these chapters provide a framework for understanding the important differences between the world of building with new materials and components, and the world of reclamation and recycling.

The remaining five chapters of the book provide guidance on the opportunities for reclamation, reuse and recycling for each of the main elements of buildings: foundations, building structure, the building envelope, enclosure and interiors, and mechanical and electrical services.

For each major element, guidance is offered on three ways in which components and materials can be reused or recycled:

- reuse in situ;
- reuse of salvaged or reconditioned products and reclaimed materials;
- the use of RCBPs.

Although each of these chapters provides guidance on their range of building elements, readers should not read them in isolation – they are intended to be read in conjunction with the general issues addressed in the first three chapters.

Finally, the reader should be aware that the book does not provide comprehensive guidance on how to salvage, reclaim and reuse components or materials from buildings being demolished. Space limitations aside, such knowledge and understanding is better dealt with by the hundreds of practical specialists whose skills and experience are as essential as they are difficult to capture in a book. As with all building design and construction, it is assumed that designers communicate and collaborate closely with appropriate specialist suppliers and contractors to ensure that their design schemes are successfully realized.

Bill Addis
February 2006

1 The World of Reclamation, Reuse and Recycling

Why do it?

Very little happens in this world unless there are good reasons for it to happen. There are three main reasons why the reclaiming and recycling of goods and materials in building construction is already happening and will grow in the years to come:

1. To reduce the impact of building construction on our environment.
2. To bring benefits to a building project, for example getting planning permission or reducing costs.
3. To improve the reputation of those engaged in building construction.

For the good of the environment

The dominant reason for reusing or recycling materials and goods is to reduce our society's impact on the environment – the world we live in (Berge, 2001). The activities of the construction industry in building new buildings and refurbishing old ones, in response to society's demand for a better standard of living, are seen to have a particularly great impact on our environment. This impact can be manifested in many ways:

- depletion of non-renewable natural resources – both minerals and fossil fuels;
- air pollution from manufacturing processes and road transport;
- degradation of the natural landscape – quarries, loss of woodland, landfill sites.

While society appreciates the improved standard of living that better buildings bring, it also sees that environmental impacts can have a detrimental effect on our overall quality of life. In recent years, this conflict has led to a growing pressure from many directions, both on and within the construction industry, to increase the reuse and recycling of goods and materials.

It is not the purpose of this book to persuade people of the need to reduce our impact on the environment, but it is worth noting that construction and demolition activities account for a large proportion of materials used and waste generated. The figures can be startling:

- In the mid-1990s in Britain the construction industry used over 250 million tonnes of crushed rock and gravel as well as nearly 3.5 million tonnes of metals, around 0.5 million tonne of polymers and nearly 4 million cubic metres of timber (Kay, 2000).
- In the late 1990s, around 10 million tonnes of post-industrial waste were generated by construction processes and around 30 million tonnes of materials arose from demolition (Biffa, 2002).
- In Britain a decade or so ago, over 3.5 billion new bricks were used each year, while around 2.5 billion were knocked down in the demolition of buildings: of these only about 140 million were salvaged and reused – the remainder were consigned to landfill (Kay, 2000).

Perhaps the most powerful statistic is that provided by assessments of the environmental footprint of mankind's activities – the area of land that would be needed to provide all our materials and energy requirements, and to deal with the disposal of waste. London, for example, would need an area of land about the size of Spain to be fully sustainable at present levels of consumption. If every country were to have the same environmental impact as Western countries currently do, we would already need more than three Earths to ensure our long-term, sustainable survival. Clearly mankind has to do something.

Figure 1.1 The environmental footprint of mankind's activities at Western levels is already greater than three Earths

Source: US Fish and Wildlife Service

The news is not all bad, however. A survey in the UK in the late 1990s revealed that nearly 2 million tonnes of materials and products were being reclaimed and reused or recycled (Kay, 2000).

Table 1.1 Selected reclaimed products and materials

Material type	Annual quantities (tonnes)
Architectural and ornamental antiques	141,000
Reclaimed timber beams and flooring	242,000
Clay bricks	457,000
Clay roof tiles	316,000
Clay and stone paving	694,000
Total	1,850,000

Source: McGrath et al, 2000

The story in many other countries is similar. In Europe, the Netherlands, Scandinavian countries and the German-speaking nations already achieve greater reuse and recycling in the construction industry than the UK. In yet other countries, while not yet widespread, there are many examples of good reuse and recycling practices.

Metals are relatively easy to separate in demolition processes – steel and aluminium can be collected using electro-magnetic methods, and typically 90 per cent or more is reclaimed and returned to the production plants where it is mixed with virgin metal. Relatively little steel is consigned to landfill and a significant proportion – from 10 per cent to over 30 per cent according to its form – is reclaimed and reused.

Other materials are more difficult to recycle and often have to be done by hand – or rather by eye, since most separation of waste materials is based on visual examination. A number of techniques have been developed to bring some automation to the process, for example, the separation of bricks using colour-recognition and plastic bottles using shape-recognition technologies.

Despite these successes, much demolition material is sent to landfill, and the numbers of landfill sites available is diminishing while taxes are increasingly being used to discourage the disposal of materials in landfill sites. In densely populated countries such as The Netherlands and Switzerland, it is already extremely expensive.

There are thus compelling reasons for trying to reduce quantities of waste materials by increasing the recycling of materials and, whenever possible, exploiting opportunities for reusing components from buildings before 'materials surplus to requirements' become simply 'waste materials'. Furthermore, shifting the balance from recycling to reclamation and reuse can reduce the reprocessing involved and hence lead to energy savings. Achieving these goals would not only reduce the growing pressure on landfill sites, it would also reduce the need to extract new raw materials from the earth. This would reduce

Table 1.2 Average end-of-life scenario data for steel in European Union

Building product made of steel	Recycling (%)	Reuse (%)	Landfill (%)
Light and heavy structural sections, mechanical and electrical services	89	10	1
Composite floor decking	81	15	4
Steel in composite cladding sandwich panels	53	37	10
Profiled cladding and roofing sections	81	15	4

Source: Durmisevic and Noort, 2003

the environmental impact of extraction processes and slow down and eventually halt the depletion of finite resources on our planet.

For the good of the project

Reclamation, reuse and recycling can benefit the building client or developer by adding value to a project, though it cannot yet be claimed that this will apply to every building project. The most common reasons, which may apply only to some parts of a building, are probably these:

- avoiding demolition and reconstruction costs by reuse in situ;
- reducing the costs of sending materials to landfill sites, for example by reusing demolition materials on site;
- getting planning permission, especially in conservation areas, by matching new construction to construction materials and methods in adjacent buildings;
- using cheaper reconditioned plant or equipment in preference to new;
- gaining credits in assessments of environmental impact that reward use of recycled materials, for example Building Research Establishment Environmental Assessment Method (BREEAM) in the UK, Leadership in Energy and Environmental Design (LEED) in the US, and sustainable construction checklists used by a growing number of local authorities in assessing planning applications;
- demonstrating a commitment by the developer and members of the project team to doing something to reduce the environmental impact of construction.

The main barriers to reclamation and recycling are unfamiliarity and inertia – being unaware of what can be done and how it can be done. The remainder of this book addresses these issues by considering the two key features of reuse and recycling:

1. There must be commitment to reclamation, reuse and recycling by the client and the entire project team.
2. The design and procurement process for using reclaimed goods and recycled materials is entirely different from normal building practice.

The world of reclamation, reuse and recycling is almost like a parallel universe that is virtually invisible to those familiar only with new building materials and components. A certain amount of background information is needed to enable project teams to overcome this unfamiliarity.

Government policies and legislation

National, regional and local governments in many countries now have policies relating to sustainable construction and these all include a commitment to minimize the waste generated and maximize the quantities of material reused and recycled. Key reasons underlying this aim are the growing realization that it cannot be sensible, in the long term, to be as wasteful of non-renewable resources as our present society often is, as well as the imminent scarcity of landfill capacity or sites, especially in small countries like the UK, Switzerland, Austria and The Netherlands.

Government policies tend to focus on two aspects of environmental impact relating to the use of materials:

1. Reducing extraction of new materials – reusing components and materials more than once brings environmental benefits in several ways. On the supply side, the demand for primary materials is reduced, as well as the resources needed to process primary materials.
2. Reducing materials sent to landfill – reusing components and materials also takes material out of the waste stream before it goes to landfill.

Achieving these goals is encouraged in the UK by a number of means and other countries employ similar measures.

Landfill tax

To reflect the need to reduce materials sent to landfill, the UK Government has imposed a landfill tax since 1996. Landfill tax on domestic waste in Britain is already UK£28 per tonne. Two rates apply for non-domestic waste: a lower rate of £2 per tonne applies to inactive (or inert) waste (as defined in the

Landfill Tax (Qualifying Material) Order 1996) and a standard landfill tax rate applies to all other taxable waste. This standard rate is set to increase annually from the current £15 per tonne up to £35 per tonne during the coming decade (Department for Environment, Food and Rural Affairs, www.defra.gov.uk).

These rates are low in comparison with other EU countries, for instance, about £50 per tonne in Germany and £100 per tonne in Holland. Holland is already committed to achieving a target of zero waste sent to landfill. It is likely that rates in the UK will rise substantially during the life of all buildings currently being designed.

The amount of construction waste and demolition waste going to landfill has reduced considerably since the introduction of landfill tax. Industry has found alternative uses for waste, but much more could be reused.

Aggregate levy

Since April 2002, an aggregate levy is charged at £1.60 per tonne and is intended to help the UK government meet its target replacement of 20 per cent primary aggregate by reclaimed and recycled materials by 2006. Like the landfill tax, this levy is likely to rise substantially in the near future.

Planning permission

A growing number of local authorities are incorporating environmental criteria when awarding planning permission to a proposed development. Authorities often publish guidance for design teams on good practice in sustainable construction as *Supplementary Planning Guidance Notes* in England, for example. Such guidance usually encourages the reuse of goods and recycling of materials. Although it is not obligatory to follow such guidance, some authorities also require an environmental or sustainability checklist to be completed and submitted with a planning application, and may impose planning conditions related to the score achieved using the checklist.

In most countries, large building projects, such as sports stadia and shopping centres, must undergo a formal environmental impact assessment, and the result of the assessment submitted in an Environmental Statement with the planning application. A requirement of the assessment of each aspect of environmental impact is that mitigation measures be proposed for how the impact will be reduced. Concerning the use of construction materials, it is now not uncommon for the reuse of goods and recycling of materials (especially construction and demolition waste) to be proposed as one of the mitigation measures.

Building assessment tools (BREEAM, LEED *etc.*)

Although not a legal requirement, the environmental impact of a growing number of buildings is being assessed using whole-building assessment methods. These are much simpler and quicker to use than calculating environmental impact from first principles. Various tools for different types of building have been developed in many countries. The assessment usually consists of awarding credits or points according to the measures taken to reduce the environmental impact of a building, for example, low energy use, use of refrigerants with zero ozone depletion potential, use of timber from sustainable forests, and so on.

Whole-building assessment tools usually award some credits for reusing materials, for example, for retaining the building fabric of an existing building, or using products with recycled content. The precise wording of the criteria for awarding such credits will help the project team choose appropriate wording for specifications that will ensure the inclusion of reused/recycled goods and materials in buildings. Some examples of the credits dealing with reuse and recycling from BREEAM, LEED and the Green Building Challenge assessment tools are given in Appendix B.

Producer responsibility

Following trends in the manufacture of cars, ships, chemicals and consumer goods, more and more manufacturing industries are being required to take greater 'producer responsibility' for the materials they use and the harm or impact they may have on our environment and society at large. In the European car industry there is already legislation forcing producers to take responsibility for their product throughout its life cycle and, hence, reduce to a minimum the materials sent to landfill. Although it is not possible to predict when or how similar legislation may be introduced into the construction industry, it is widely agreed that it is a question of 'when', not 'if', and many people are talking of a time scale of 20 to 30 years.

This driver is also likely to move producers towards product service systems where they maintain ownership of the product and lease it to the consumer, maintaining, repairing and upgrading as necessary, and ultimately taking it back. This has the added advantage for the producer of knowing the state of the product at all times, and being able to reduce long-term risk of future product development plans.

For the good of your organization

The public, the media, civil servants and politicians are now all aware of the damage that the environment is suffering as a result of many everyday activities. A consequence of this growing awareness is that people are becoming less tolerant of those who continue to undertake such activities without trying to reduce their impact on the environment. In various ways and to varying degrees there is a growing competitiveness among many organizations to appear more concerned for the environment than their rivals. An organization's reputation or track record in environmental matters may now have an impact on many general aspects of business:

- environmental record compared with competitors;
- annual reports to shareholders;
- listing of shares on environmental or sustainable indices (e.g. FTSE-4-Good in UK, Dow Jones Sustainability Index in US);
- attracting new employees (especially young ones).

The reputation of organizations in the construction industry is likely to affect their ability to get work or sell their goods or services:

- being invited to tender for work (especially government contracts);
- being invited to join project teams;
- inclusion on lists of preferred suppliers.

Being seen to take seriously the reclamation and reuse of goods and recycling of materials in a construction project can be a highly visible way of demonstrating a commitment to improving our environment or, at least, reducing damage done to it.

Reclamation, reuse and recycling are not new ideas

The reuse of building elements and recycling of materials are not new ideas. On the contrary, until the 19th century it was the norm throughout the world. Today, it is still widely practised in all poorly industrialized countries:

- From the earliest days of large-scale masonry construction in ancient Egypt, Greece and Rome, large dressed stones have been reused many times as buildings were destroyed by earthquakes or in war, or simply fell into disrepair. The manpower needed and the cost of reusing such stones was much less than hewing new stones from quarries that might be many hundreds of kilometres distant.
- Likewise with iron. Hardly any iron from Roman times has been found, yet they used many millions of tonnes of wrought iron in their buildings between around 100BC and AD500. Nearly all of it has been reclaimed, reused or recycled, both in the building industry and to make machines and weapons.
- The Roman building engineer Vitruvius, writing in around 25BC, advises that the strongest walls are those made using old fired-clay roofing tiles since only the best quality tiles would have survived the ravages of rains, winds and frosts. He also mentions that murals painted on brick walls could be cut out with their brick backing, packed in a timber frame for transportation and incorporated into another building.
- Most medieval cathedrals were constructed on the sites of earlier churches and, wherever possible, they would incorporate both their foundations and crypt below ground level, and all the masonry of the earlier building above ground.
- Since the end of the 19th century most steel has been manufactured using a proportion of scrap (recycled) iron or steel in order to reduce

its cost. Through much of the 20th century this fact was not widely publicized as users of steel did not like the idea that they were buying second-hand material, even though the engineering properties and quality of the steel were in no way affected. More recently the aluminium industry has made much of high recycling rates in order to deflect criticism for the large quantities of energy used in the production of virgin aluminium.

- The last half-century has seen a growing number of architectural salvage firms stocking goods ranging from low-value doors or paving stones to high-value architectural ironmongery and ornaments. These have specialized mainly in high-value goods – either inherently high-value, such as wrought-iron gates, or those whose value arises from their scarcity, such as Victorian fireplaces or medieval roof tiles and bricks.
- Small-scale builders with a need for just one or two doors or steel beams have long sourced their goods in various types of salvage yard. And at a domestic level, which of us has not removed some item from a waste skip outside a house where builders are at work?

Experience in other manufacturing industries

The issue of recycling and reuse is already being addressed in several manufacturing industries. In the European Union, this is being achieved by means of a legal framework based on the notion of 'producer responsibility'. A series of proposed Directives and national producer responsibility legislation is evolving and already affects the automotive industry, electronic and electrical industries and industries connected with manufacturing and using packaging. The long lead time involved in developing new cars (up to a decade) means that car firms began addressing the end-of-life issues long before the legislation came into force:

- Packaging Directive (1994);
- End-of-life Vehicles (ELVs) Directive (2000);
- Waste Electrical and Electronic Equipment (WEEE) Directive (2003);
- Restriction of Hazardous Substances (ROHS) (2003).

It is worth noting developments in other industries because they are likely to serve as models and precedents for when (not if) similar legislation is developed for the construction industry, for instance:

- how deconstruction and recycling has been embedded in the infrastructure of the industries;
- how the structure of the industry and firms within it have changed;
- the means of facilitating deconstruction, reuse and recycling;
- the type of legislative framework that is likely to be introduced in the construction industry during the next half-century.

The automobile industry

A series of articles under the End-of-Life Vehicles Directive cover:

- bans on hazardous substances and materials (Article 4);
- economic operators to set up collection systems (Article 5);
- stipulations on storage and treatment methods for ELVs (Article 6);
- requirements and targets for reuse and recovery (with a preference for recycling) (Article 7).

By 2006 a minimum of 85 per cent by average weight per vehicle and year must be reused, recycled or recovered – 80 per cent minimum reused or recycled and a maximum of 5 per cent energy recovered:

- for pre-1980 vehicles the figures are 75 per cent and 70 per cent, respectively;
- from 2015 it will be a minimum of 95 per cent reused, recycled and recovered, of which a minimum of 85 per cent must be reused or recycled.

Box 1.1 Options for end-of-life of vehicles by Honda

Faced with the forthcoming End-of-Life Vehicle Directive, Honda considered four possible options that would be available to them:

Option 1

A contractual scheme, whereby producers are responsible for putting in place collection schemes for their own vehicle components. This option would provide flexibility for producers, who could have individual contracts or participate in a collective network.

Option 2

A tonnage target scheme, where vehicles are delivered to treatment facilities with a permit. The producer would have to meet all or a significant part of the cost of take-back and treatment for negative value materials. In addition, each producer would be required to meet tonnage recycling targets, which might be set according to each producer's market share. Targets would be met by purchasing evidence notes.

Option 3

A hybrid scheme:

- For existing cars, tonnage targets would be set based on the current market share and evidence notes purchased.
- For new vehicles, producers would be required to pay bonds into a fund for each vehicle sold.

Option 4

The last owner would be required to deliver an ELV to an authorized dismantler or shredder. The dismantler would accept the car free of charge and the shredder would be obligated to accept the vehicle as long as it was complete. This option would be funded by a levy on every new vehicle sold and reported.

Electrical and electronic equipment

The production of electrical and electronic equipment is one of the fastest growing domains of manufacturing industry in the Western world. Technological innovation and market expansion accelerate the replacement process and new applications of electrical and electronic equipment are increasing significantly. Therefore the resulting rapid growth of waste from electrical and electronic equipment (WEEE) is of concern. Figures show that the growth of WEEE is about three times higher than the growth of the average municipal waste. Recent estimates are that in the UK alone we already have 90 million redundant mobile phones sitting in drawers. These are estimated to contain precious metals worth over £20 million.

The hazardous contents of WEEE cause high concern when these products become waste, since these are not separately collected and pre-treated and end up in municipal landfill sites where appropriate measures for preventing the hazardous substances from entering the environment are missing.

As a response to these concerns, the European Commission adopted in June 2000 two proposals for directives, one on WEEE and one on the restriction of the use of certain hazardous substances in electrical and electronic equipment. Non-binding targets for collection from households are set at 4kg per inhabitant per year and a series of recovery targets for items including appliances, consumer equipment and IT and telecom has been proposed.

The proposed time scales for implementation are as follows:

- Directives commenced – January 2004;
- producer responsibility – January 2005;
- meet recycling targets – mid-2006.

The recovery and reuse targets currently set are shown in Table 1.3.

Table 1.3 Recovery and reuse targets

Category	Recovery	Reuse
Large household	90%	75%
IT and consumer	85%	65%
Others	70%	50%

Source: DEFRA, www.defra.gov.uk

Under ROHS, substance 'substitutions' are also listed. The following materials have a proposed phase-out date of January 2008:

- lead;

- mercury;
- cadmium;
- hexavalent chromium;
- brominated flame retardants (PBB and PBDE).

Exemptions are available if there is no substitute, or when the proposed substitute is worse for the environment or health.

The lesson for designers is that it will become increasingly important to be aware of the materials considered to be hazardous and if possible, materials that might, in the future, be deemed to be hazardous.

Box 1.2 Altran Technologies: Design for Disassembly

The electronics industry, perceiving an upturn in volume and stringency of legislation, has started to prepare. AltranTechnologies UK, in collaboration with the aerospace industry, has developed a web-based tool for the electronics industry to aid them in Design for Disassembly. The tool is currently used on 100 per cent of Motorola products and has the potential to be adapted for more applications. The tool follows a series of steps following the substance or element through materials, concept design and review and detailed design stages.

For example at the substance stage, general data are available on:

- composition;
- physical properties;
- applications;
- abbreviations and specifications.

At each stage, detailed environmental information is also available on:

- recyclability;
- maximum concentrations in materials to allow recycling;
- composition;
- reporting requirements;
- toxicity;
- global legislation.

Similar categories and thought processes are likely to become relevant to the buildings and the equipment installed in them.

It is likely that the WEEE Directive applies to certain equipment such as white goods, lighting equipment, IT equipment and medical equipment systems that may be installed in buildings. However, it is not yet clear how this will affect their end-of-life in a building, nor is it an issue that building designers can influence.

The basic concepts of reclamation, reuse and recycling

The life cycle of materials

The very idea of 'waste' is one that belongs to the throwaway society, not to the reuse and recycling society. Early endeavours to reform our ideas about the life cycle of materials tended to focus on the 'waste problem', in other words, what could be done to use the mountains of waste instead of sending it to landfill. This was the focus of much guidance in the 1990s. Today, many people have realized that the very idea of waste belongs to the old way of thinking in which waste is seen as a problem. Gradually more and more people are now looking at things differently – looking at the life cycle of materials and the various 'materials streams' that can be traced through manufacturing processes in the construction industry.

The most successful way of dealing with waste is not to produce it. Rather than seeing materials at the end of their first life as a problem, they can be looked at as an opportunity. The concepts of 'waste', 'reuse' and 'recycling' are best understood in the context of the life cycle of materials. Recent and many current practices have tended to involve a linear flow of materials from 'cradle to grave' characteristic of the throwaway society (Figure 1.2).

The ideal situation to which we should aspire if we are to avoid waste is a circular or closed-loop life cycle similar to that found in natural ecosystems (Figure 1.3).

In practice, we are not likely to achieve this ideal in the foreseeable future for most building construction. However, an important contribution that people in the building industry can make to approaching it is to design and construct buildings differently. They could be designed in ways that make reuse and recycling happen (Addis and Schouten, 2004).

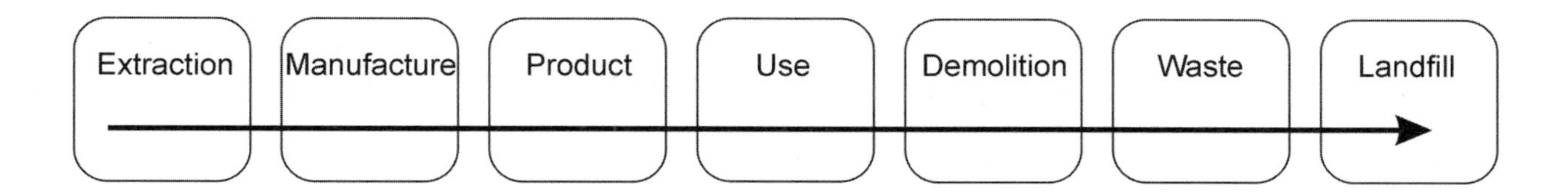

Figure 1.2 The linear life cycle of materials and goods

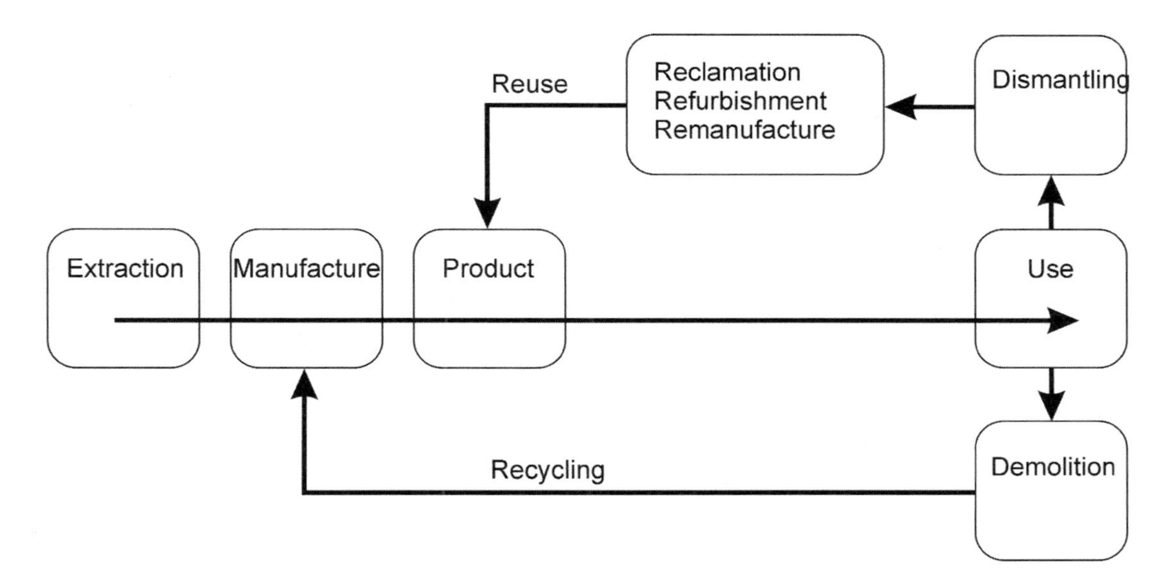

Figure 1.3 A closed-loop life cycle for materials

The Delft Ladder

The Delft Ladder (Figure 1.4 and Box 1.3) was devised as a way of representing in a diagram the various possible stages of the life cycle of materials (te Dorsthorst et al, 2000). The ten stages in the Delft Ladder represent the stages of the life cycle of materials and components at which the designer can take action to ensure they are used at the highest possible level in their life cycle, for as long as possible. By this means the degradation of materials towards landfill is prevented or slowed down. When making design choices, the designer can consider, *in sequence*, each stage of the life cycle:

1. *Prevention* – how can waste be avoided by design? For instance reducing waste by minimizing material use or eliminating components.
2. *Object renovation* – how can waste be avoided by prolonging an element's life in use? For instance through maintenance.
3. *Element reuse* – how can an element be reused after removal? For instance refurbishing a kitchen sink.
4. *Material reuse* – how can materials be reused after removal? For instance reconditioning bricks for reuse.

5 *Useful application* – how can materials or elements be recycled or reused in new applications? For example using crushed bricks as hardcore or using a structural steel beam in a temporary works structure.

How to reuse or recycle components and materials

There are three ways in which materials and products used in buildings can be prevented from becoming useless and from ultimately being sent to landfill sites.

Reusing in situ

The least amount of intervention is likely to be required if a building component or even a whole building can continue to be used, or be reused, in its existing location. This will avoid the creation of much demolition waste that will have to be dealt with, as well as the need for large quantities of new building materials.

Effective cleaning, repair and maintenance, servicing and refurbishment will all help to prolong the life of a building and its various elements. While these activities are not really within the scope of this book, they are important in helping to achieve the ultimate goal.

When a building is to be redeveloped for a new use, the easily removable items, such as light partition walls, building services and various fixtures and fittings, may be stripped out and replaced. However, it is likely that the structure and foundations of the building and much of the building envelope can be retained. To do this it would be necessary to under-

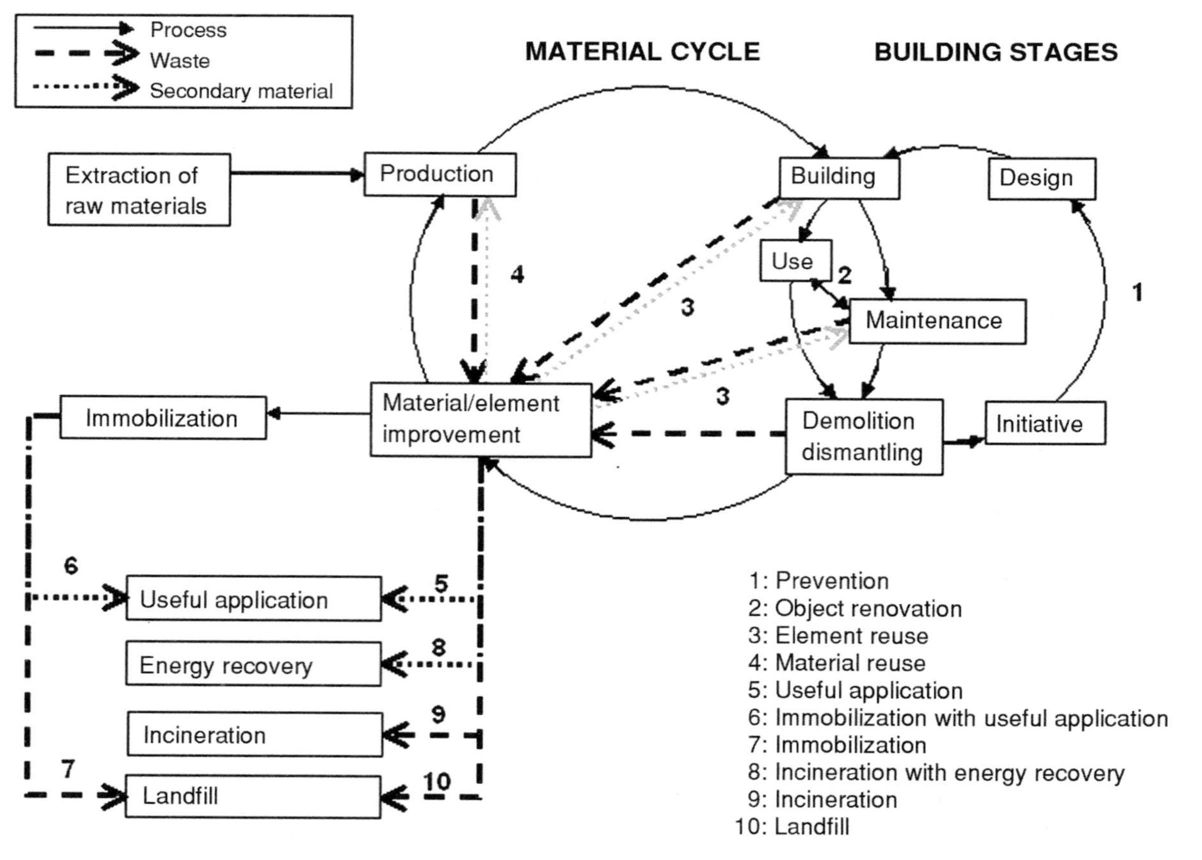

Figure 1.4 The Delft Ladder showing life-cycle material flows

Box 1.3 Key to the Delft Ladder

Processes of the Delft Ladder

1 Prevention – waste can be avoided by devising a building system that allows dismantling and reassembly, or by designing a component that allows the materials to be fully recycled.
2 Object renovation – avoiding demolition by renovating a building and its components to keep them in continued use.
3 Element reuse – elements removed from a building can be improved by maintenance, refurbishment or reconditioning and reused for their original purpose in a new situation.
4 Material reuse (recycling) – waste materials from production processes can be collected and improved (for example by cleaning) to make them suitable for returning to the production process.
   Following processes 3 or 4, if the material or element cannot be reused for their original purposes, they can be improved to enable their use in a new application.
5 Useful new application – a material, element or component can be used in a different situation, perhaps with a lower performance specification ('down-cycling').
6 Immobilization with useful application – a potentially harmful material can be rendered harmless when used as a raw material for a new component (e.g. the use of pulverized fuel ash in concrete).
7 Immobilization – a potentially harmful material can be rendered harmless before sending to landfill.
8 Incineration with energy recovery – combustible materials are burnt and the energy liberated collected for use.
9 Incineration – combustible materials are burnt and, though not providing useful energy, are not sent to landfill.
10 Landfill – the final destination of materials if no alternative use can be found.

take a careful assessment of their condition and their suitability for the new building use. A certain degree of repair and refurbishment may be necessary, but would amount to less work than demolition and reconstruction (see Figures 5.7 and 5.8).

Reusing salvaged or reconditioned products and reclaimed materials

If the major items of a building cannot be reused in their original location, it may be possible to reuse them in a new location. The most dramatic way this can occur is by bodily moving an entire building (see Figures 5.4–5.6).

It is more usual for elements of a building to be removed during refurbishment or demolition and reused in another building. Normally the item will need to be worked upon in some way to bring it up to the standard required for it to be reused. Different products are likely to require different levels of treatment to achieve this, for example:

- A steel beam may need to be cleaned, cut to length, prepared for new end-connections and corrosion protected.
- A hot-water radiator may need to be cleaned inside and out, have new valves fitted and be pressure-tested to detect any leaks.
- An electric motor or the chiller from an air-conditioning system may be returned to the factory where it was made, stripped down and rebuilt with the original, reconditioned or new components, as required, to achieve the desired performance. It would then be supplied with a warranty.

Products or materials thus brought back into use are called 'reclaimed' products or materials.

Using recycled materials

Recycling refers to the use of waste materials to make new products, usually different from the products in which the materials were used during their previous life. Typical examples include:

- Chipboard made from sawdust collected in sawmills or timber reclaimed from building demolition and suitably processed.
- Concrete made using recycled aggregate (crushed concrete from building demolition).
- Plastic drainpipes made from the reclaimed plastic drinks bottles.
- Formwork for concrete piles and columns using cardboard tubes made from recycled paper.

- Thermal or acoustic insulation made from old newspapers.
- Acoustic isolation rubber matting made from old car tyres and car window seals.

There are three distinct sources of recycled materials. Materials collected at the source of extraction (a mine, quarry or forest) as waste products from the manufacture of a primary material are called *secondary materials*. Materials that are collected in factories as a waste product from manufacturing processes (such as sawdust and timber off-cuts) are called *'post-industrial' by-products* or *waste*. Materials collected after they have served a useful purpose in a product (such as a car tyre) are called *'post-consumer' waste*.

Given a choice, it is more environmentally beneficial to use post-industrial waste as the material has not progressed as far along the cradle-to-grave life cycle as post-consumer waste. Similarly, it is better from an environmental point of view to reuse components or equipment rather than to use recycled materials that have already progressed further along the material life cycle.

Many products are made using a proportion of recycled material mixed with virgin material of either the same or a different kind. Such products are often labelled giving the percentage of recycled material, for example plasterboard made with 50 per cent post-industrial waste gypsum, or paving tiles made using 80 per cent post-consumer bottle glass.

Many manufacturers are now making products using a proportion of recycled materials and, especially in the US, these are generally referred to as recycled-content building products. This phrase has been adopted in this book, usually abbreviated as RCBP.

It can be especially important to know the recycled content of products if the environmental impact of a building is being assessed using an environmental assessment method that awards credits according to the percentage of the building that is reused or recycled (see Appendix B).

A hierarchy of reuse and recycling

When considering the options for reuse or recycling available for a project, it will be useful to devise a hierarchy that will help decide which option will be most appropriate from the environmental point of view. This approach has been taken by a number of authors (e.g. Anink et al, 1996; Woolley et al, 1997; Woolley and Kimmins, 2000) and is sometimes called an 'environmental preference method'.

In the case of reuse and recycling, the options to be considered by the design team need to reflect the practices of the reclamation and salvage industries, since these control the goods and materials that will be available to the design team. One example of such an approach is the 'recycling protocol' devised by the UK salvage firm Salvo based on the hierarchy of options: reuse, reclaim, recycle, destroy (Box 1.4).

The reuse and recycling market place

The materials and building components and equipment that are available today for reuse and recycling are governed by a number of factors:

- how buildings were constructed in the past;
- the durability of buildings and their elements;
- current methods of deconstruction and demolition (see Appendix A);
- the demand for reclaimed goods and materials.

The market place is developing. Thirty years or so ago recycling was just starting to be taken seriously in Scandinavia, The Netherlands and German-speaking countries. Thirty years ago there were few architectural salvage yards; today there are many.

The market place for reused/recycled materials and building products in the future will be influenced strongly by how buildings are designed and constructed today. In fact, examples of buildings or building elements designed to facilitate deconstruction for reuse and recycling at the end of their life are very rare indeed, and usually found to be so almost by accident because they were designed as prefabricated components intended for rapid on-site assembly. Design for deconstruction will help the market for reused goods and recycled materials to develop more rapidly in the future.

Reclamation, reuse and recycling seen from different viewpoints

The relative importance of different issues involved in reclamation, reuse and recycling depends on

Box 1.4 A recycling protocol devised by Salvo

A recycling protocol with respect to the built environment gives the following priority list for what to do with old architectural items. 'Architectural antiques' are defined as manufactured items usually with a degree of skilled labour involved, such as carved items, doors and fireplaces, and 'reclaimed building materials' as the basic building components such as bricks and timber beams.

1. Reuse a building without demolition or alteration. If this is not possible then:
2. Reclaim components in as intact a way as possible by:
   - relocating entire buildings;
   - reusing façades and structural elements;
   - reclaiming whole features such as windows with their surrounds, shutters and window furniture;
   - dismantling and reclaiming the individual items that were used to assemble a building.

   If reclamation is not possible then:
3. Recycle and remanufacture a new product:
   - Reclaimed wood can be recycled to make furniture, floorboards or even blockboards.
   - Concrete can be crushed to make recycled concrete aggregate.
   - Plastics can be remanufactured into new plastics products like polythene bags.

   If recycling is not an option then:
4. Beneficially destroy and recover energy:
   - Scrap wood and other carbon-based products can be used to fuel power plants, or for local heating or cooking.
   - Methane can be recovered from landfill sites where carbon-based demolition waste has been tipped.

Source: www.salvoweb.com

one's point of view – one man's waste problem is another man's commercial opportunity. The last decade or so has seen a gradual shift from seeing reclamation and recycling as an issue for the waste industry to one that is relevant to everyone concerned with a building during its entire life cycle.

The waste industry (Figure 1.5)

By and large, materials and goods that have entered the waste industry will be destined for recycling and not for reuse. They will already be seen as 'a problem' but one that can be turned into an opportunity. The large numbers of scrap car tyres and plastic liquid containers have stimulated many entrepreneurs to look for products that can be made using recycled materials.

The waste industry sees recycling according to the material(s) involved, which is not how the building designer usually looks at components for use in buildings.

The demolition contractor (Figure 1.6)

The size and growth of the reuse and recycling market behave according to the familiar laws of supply and demand. As is often the case in the economy, different players are brought together in various ways and the market grows only when everyone gains something. The market place looks different from various points of view.

The easiest and quickest option for the demolition contractor is to demolish a building as quickly as possible and make as much money as possible by selling the products and materials, assuming they own them and there is a market for selling them. In practice such markets are extremely volatile – one week there may be a good price for reclaimed timber, while the next the contractor will have to pay to get the timber taken away and burned or sent to landfill. The demolition contractor will always seek to minimize the quantities of materials to be sent to landfill because of the cost – a cost that is now growing annually in most countries and already so high in some countries that landfill is no longer a feasible option.

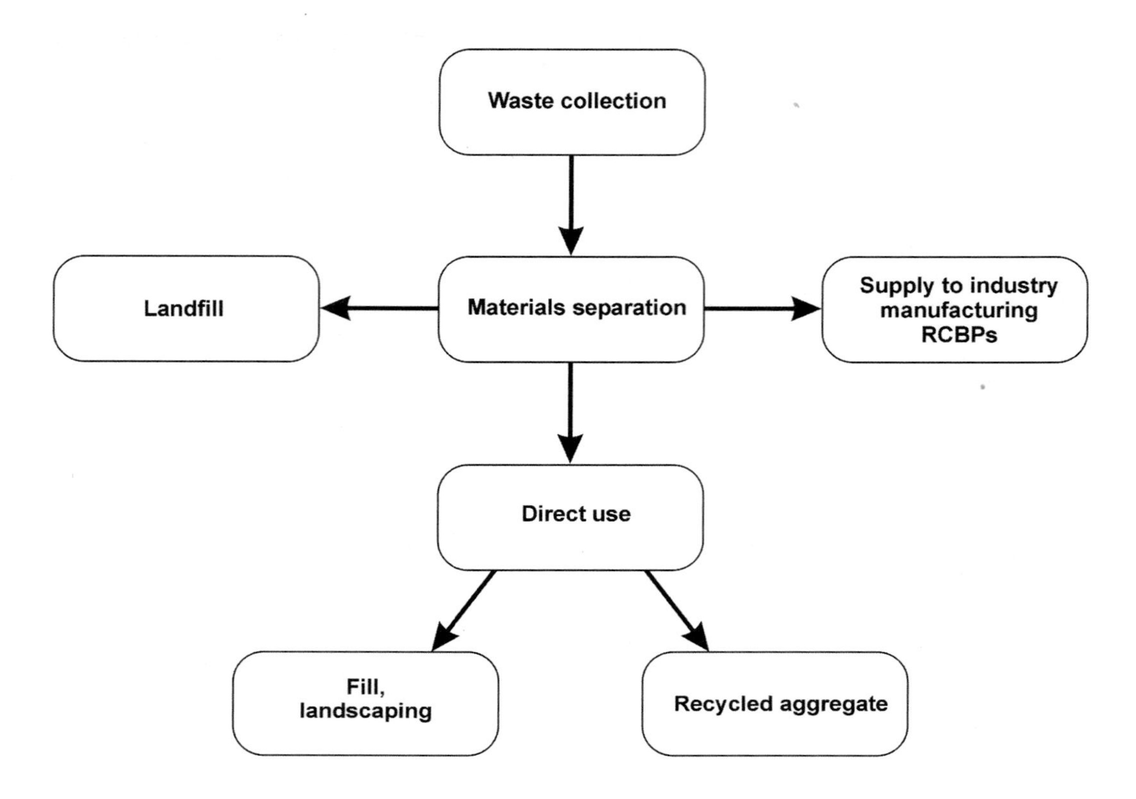

Figure 1.5 Reclamation, reuse and recycling: The waste industry view

The demolition contractor will salvage or reclaim those items and materials than can be easily removed from the building and which have a sufficient value to cover the additional effort and care needed to keep them in a suitable condition (e.g. architectural ironmongery and most metals).

Generally speaking a demolition contractor will not salvage items if there is little likelihood of a buyer coming forward quickly, thus keeping storage costs to a minimum. This will be especially true if great care is needed when removing the items from the building to ensure they will be reusable, for example cladding panels.

Manufacturers of RCBPs (Figure 1.7)

At the end of the chain in the recycling of materials are the manufacturers who use recycled materials in the manufacture of products. Their primary commercial task will be the same as all producers – to persuade the potential buyer of the unique selling point of their product.

The salvage industry (Figure 1.8)

In one sense the salvage industry – the purchase of goods deemed to be waste by their owner by someone who recognizes there are potential buyers who value the goods more highly than their 'waste value' – has been in operation for all history, alongside the market for antiques and other second-hand goods. Nevertheless, it has only developed as a significant sector of the construction industry during the last 30 years or so, that is since people began to challenge the widespread destruction of many old buildings in the name of progress and improving our old towns that occurred immediately after World War II.

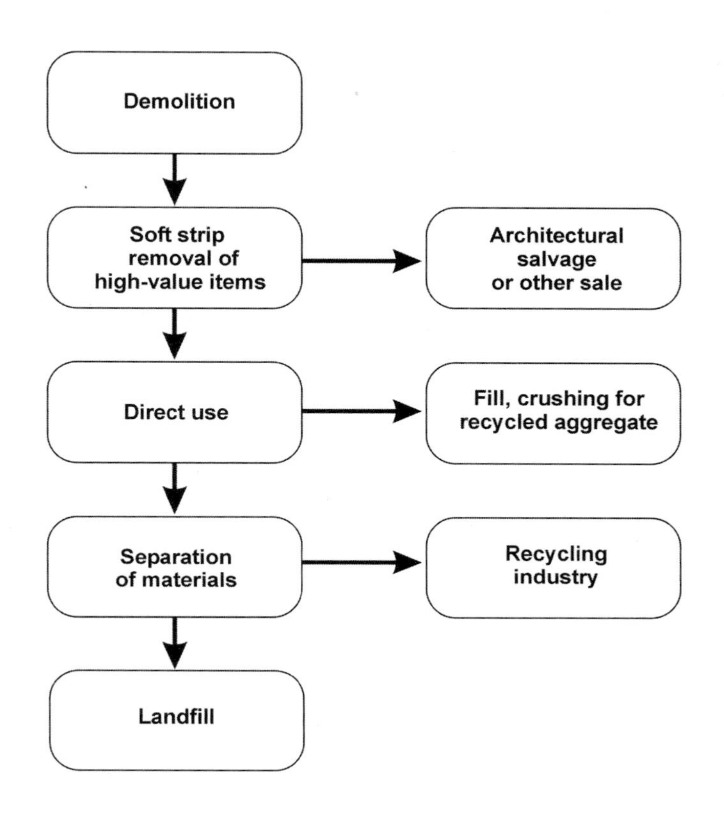

Figure 1.6 Reclamation, reuse and recycling: The demolition industry view

The way the salvage market can work is no better illustrated than by the following tale published by the UK salvage firm, Salvo:

> An example occurred in 1997 where a balustrade and supporting columns from an Edwardian cinema was thrown into a skip in London, and was spotted by two antique dealers who happened to be passing. By the time they had parked and walked back to the skip, it had been taken to a transfer station. They tracked it down and offered £150 for the item, which they sold the next day to an architectural antiques dealer from Wales, who happened to be in London for the day, for £1500. He sold it again within a week to a pub-fitter in Leeds for £5000, who restored it, thus providing employment, and fitted it into a West Yorkshire pub. Had this rescue not occurred, the chances are that the balcony would not have been landfilled, but would have been sold for £10 as scrap, and resmelted, probably in Spain or the Far East (Kay, 2000).

The client and design team (Figure 1.9)

The client and designer will have clear reasons why they are intending to reuse components and equipment, and use RCBPs in their building. At the whole-building level they may be wanting to achieve a certain percentage of reused or recycled materials. At the level of building element, they are likely to be focusing mainly on the possible sources of reclaimed components and RCBPs as alternatives to the more familiar suppliers of new items. With this in mind, they are likely to have to engage with the demolition industry if they wish to reuse components that will require special care in their removal to ensure they can be reused. Otherwise the designers will approach the design of a building using reclaimed components and RCBPs much as they approach an ordinary building – element by element.

Client and designers would need to make a number of choices about which building elements to use in situ, which to reuse from another location, which would be RCBPs and, if these were not technically possible or economically viable, which would be new items.

The building itself (Figure 1.10)

It is only seldom that a single owner has an interest in a building from its construction to its demolition. It is this fact that can conspire against the likelihood of closed-loop thinking being imminently applied to buildings.

The ideal scenario is that buildings are designed and built in ways that make them easy to dismantle to facilitate the reuse of components and equipment or, if this is impractical, that all the materials are joined together in ways that allows them to be easily separable, which will make their recycling much more likely. If this happens, many more building components will be available that can be reused or are made with a high recycled content. Designers could then use such components in a new building that would also be designed with dismantling in mind, and so on.

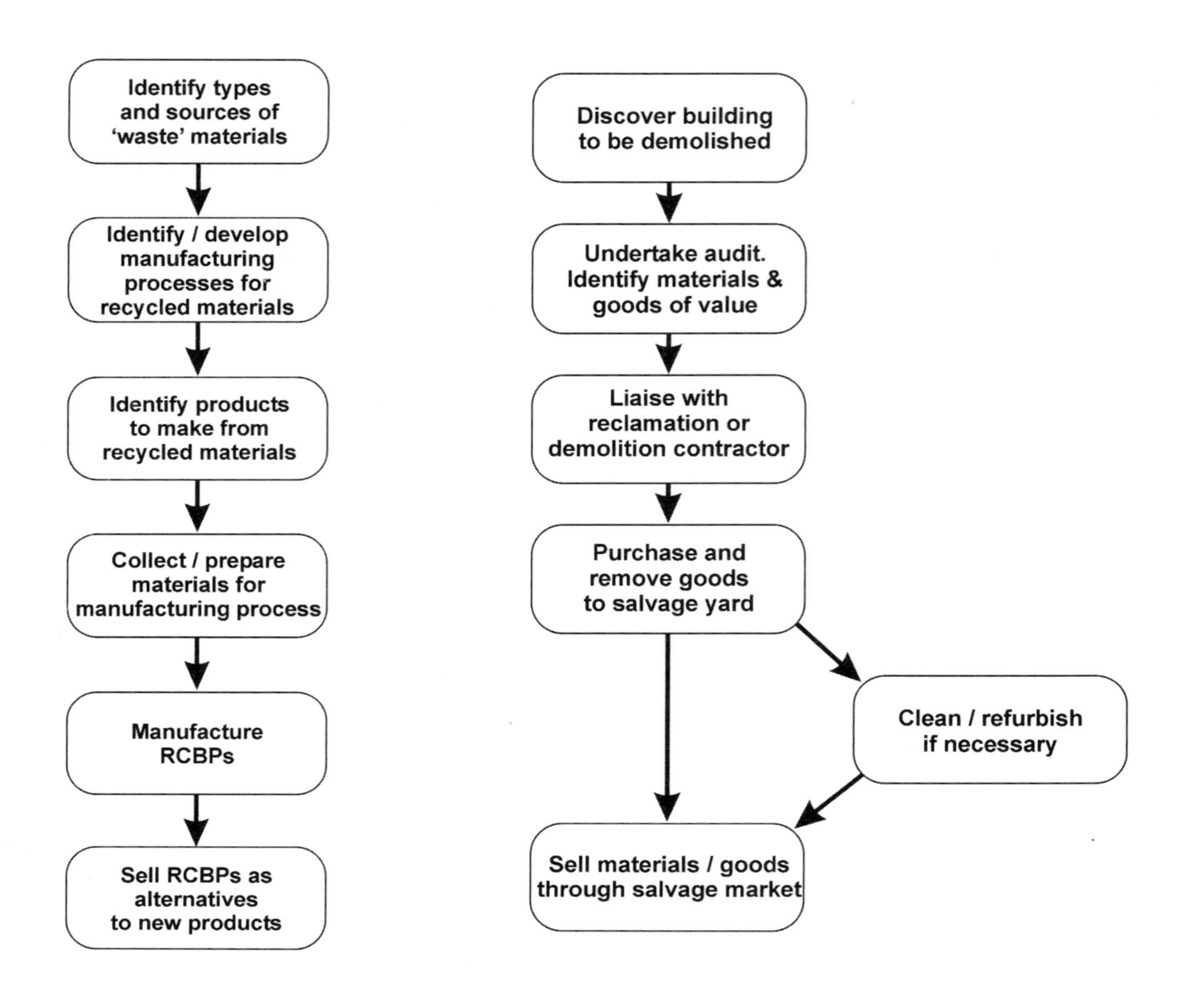

Figure 1.7 Reclamation, reuse and recycling: The recycling and manufacturing industries view

Figure 1.8 Reclamation, reuse and recycling: The salvage industry view

This would be the truly virtuous circle (Addis and Schouten, 2004).

It is perhaps worth considering for a moment what the ideal reuse/recycling building project would perhaps be:

- a building made from a child's construction kit such as Konnex, Lego or Meccano;
- a tent used by nomads, campers or circus performers;
- a travelling theatre or stage for performances by pop groups;
- temporary accommodation on a construction site, perhaps four storeys high and a dozen units long;

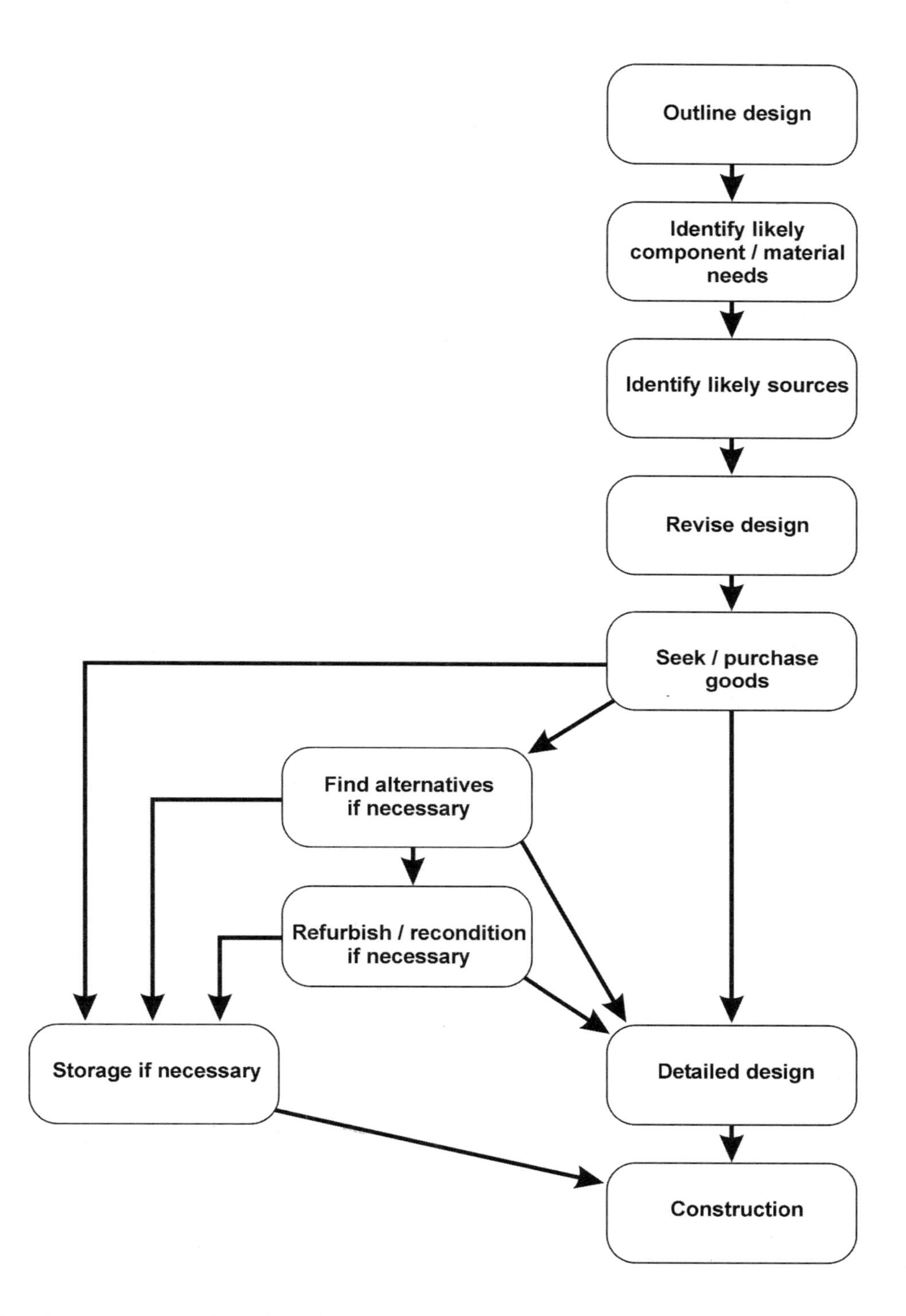

Figure 1.9 Reclamation, reuse and recycling: The design team and client view

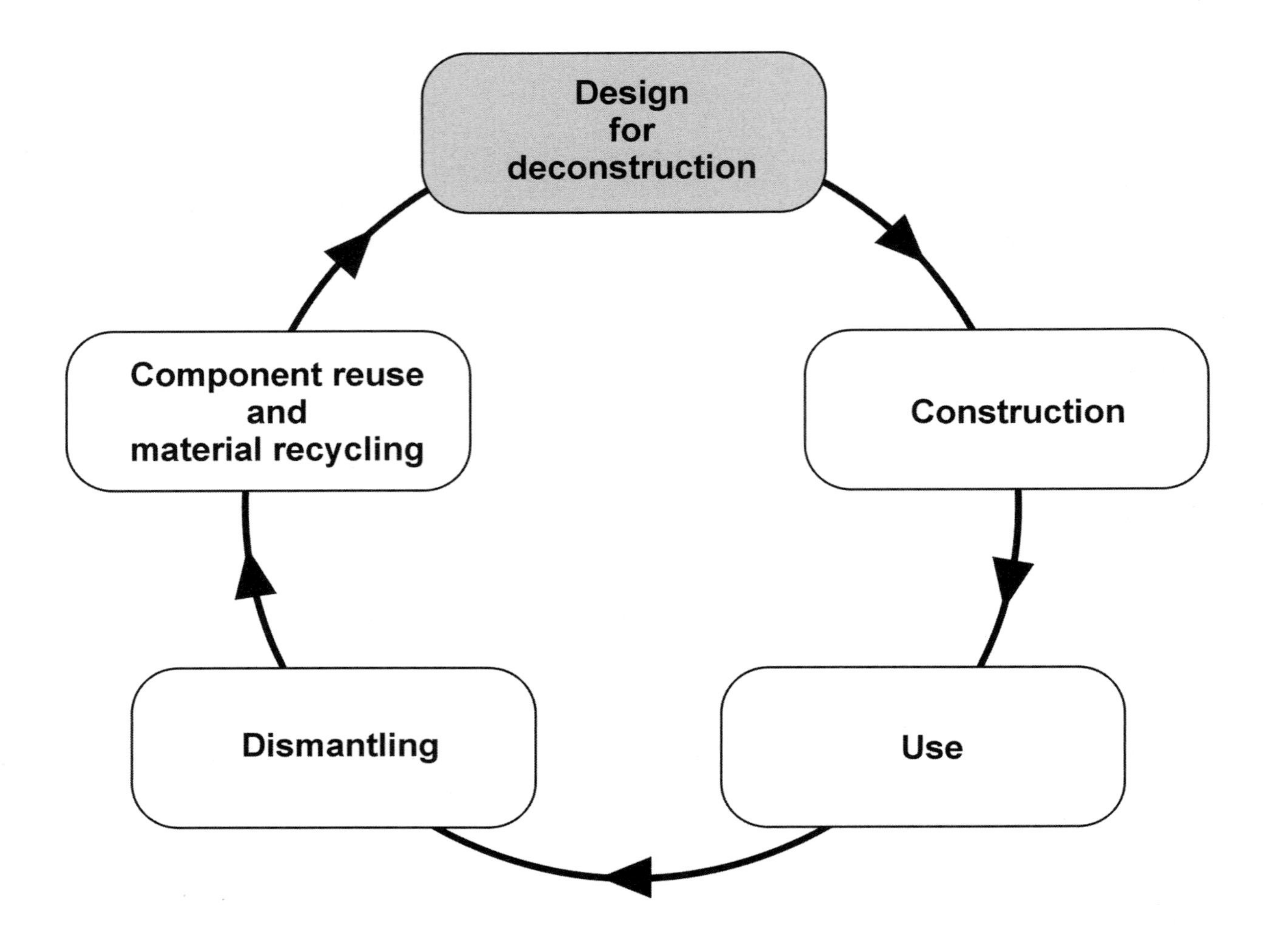

Figure 1.10 The virtuous circle: Design for deconstruction and reuse

Supply

Existing buildings

There is clearly a large supply of existing buildings that may be suitable for reuse as a whole or in part, or may be a suitable source of building products and materials that could be reclaimed and reconditioned for subsequent reuse or recycling. A client will need to take advice from specialists able to advise on the potential for reusing the entire building, for example:

- Load capacity of the structure and existing foundations, and potential for upgrading if necessary.
- Feasibility of introducing new stairs, lifts and services cores and risers.
- Feasibility of introducing modern building services and their vertical and/or horizontal distribution.
- Heritage issues concerning alterations, materials and the appearance of the building envelope.
- Likely costs of undertaking the above work.

Similarly, if a building has been identified that is due for demolition, waste and salvage specialists should be sought and consulted for advice, including for example:

- An inventory of the materials, components, plant and equipment, and an assessment of their potential for economic removal for reclamation and reuse, or for recycling (see, for example, the UK Institution of Civil Engineers' Demolition Protocol (ICE, 2004)).
- Advice on briefing demolition contractors to ensure minimum damage to components scheduled for reuse. The UK's National Green Specification, for example, calls for demolition contractors to indicate what is to be reclaimed and produce a method statement indicating how the goods will be extracted in good condition, palletized and protected during transportation and storage.
- Advice on duration and costs of demolition involving careful deconstruction and dismantling compared to normal demolition.
- Advice on the market value of any goods not needed for immediate reuse but with the potential for sale into the architectural salvage marketplace.

With reuse in mind it may be appropriate for a client to buy a vacant building and take control of its deconstruction, rather than acquiring a site after demolition has been completed. This will make it easier to benefit from the opportunity to reuse items and demolition material on site, so reducing the costs of waste removal and landfill, the costs of buying goods and materials (such as fill material for foundation backfill, service trenches and landscaping) and minimizing the environmental impact of transport.

Reclaimed goods and materials

Goods and materials with a high potential for being reclaimed and reused generally find their way into the architectural salvage market. They are recognized as being worth salvaging by demolition contractors, removed from a building and stored in an architectural salvage yard until a buyer is found or finds them.

Some goods or materials will have been little more than cleaned prior to going on display at an architectural salvage yard almost ready for immediate reuse (e.g. bricks, tiles). Some have additional value if they are left uncleaned – lichens, soot and paint can add character and help blend into an existing building and streetscapes. Others will be sold as seen, and refurbished by the buyer (e.g. window frames and doors, iron radiators). More valuable items are likely to be fully refurbished by the salvage firm to increase their sale value (e.g. cast-iron fire places, wrought-iron gates).

Other goods capable of being reclaimed and reused may not be salvaged simply because there is currently no demand for them, or they may be inconvenient or costly to store (e.g. cladding panels, air-conditioning ducts).

The availability of reclaimed goods and materials in the market place thus depends on a number of factors:

- Can they be easily removed from a building being demolished or dismantled without damage?
- Are they inherently valuable (based on original cost)? For example a marble fireplace.
- Are they valuable due to their scarcity? For example 16th-century roof tiles or bricks.
- Does there already exist an infrastructure dealing in the goods? For example architectural salvage.
- Are the goods easy to transport and store without damage until a buyer can be found?
- Are the goods in demand?
- Are the goods likely to have a useful life remaining when removed from a building?
- Are they available in the quantities that people may require?
- Is it easy to assess the condition of the goods or materials for their potential reuse, and to assess the likely useful life remaining?
- Can they be easily refurbished to restore their condition sufficiently for reuse?
- Can a suitable organization be identified that will test a product and provide a warranty adequate to meet the building designer/contractor's needs (both technical performance and for insurance purposes)?

Reconditioned products for reuse

Apart from goods and products that can be refurbished or reconditioned in architectural salvage yards or by builders, many products used in buildings can be returned to specialist firms or the original manufacturers for reconditioning. This is likely to involve stripping down or dismantling, replacing worn parts, rebuilding the product and testing it to ensure its satisfactory performance. Such goods would usually be available with a warranty from the firm undertaking the reconditioning.

The availability of such products/materials in the market place will depend on a number of factors:

- Can the product be removed from a building and returned in a suitable condition to a firm that will undertake reconditioning, including performance testing if necessary? Are the products valuable enough to make reconditioning viable?
- Is the original manufacturer still in business, either to supply spare parts, or to provide a reconditioning service on demand, or to undertake reconditioning as a core service?
- Are specialist firms (not original manufacturers) available to undertake reconditioning (for example of metal windows)?
- Can the performance of reconditioned goods be specified in ways that meet the needs of building designers and contractors?

Recycled-content building products

A small but growing number of firms is focusing on the manufacture of products that use recycled materials. Their further growth will depend on their success in market penetration in competition with the manufacturers who use virgin materials, and the demand for products with recycled content. One web-based database already lists over 500 RCBPs and some 1400 recycled materials that are available (NGS, 2004).

Demand

Ultimately the success of the various markets for existing buildings, reclaimed materials and goods and RCBPs depends on there being sufficient demand. Depending on the type of reuse or recycling the demand will come from one of the following:

- government department procurement directives;
- the building owner/developer/client;
- members of the design team;
- persons who write specifications;
- main or specialist contractors.

These people will only specify RCBPs or recycled materials if there are good reasons to do so. As always, the reason foremost in many people's minds will be cost. There are already many examples of reclaimed or reconditioned goods, RCBPs and recycled materials that are cheaper than those made with virgin materials.

Often, however, it is not cost alone that affects decisions about building design and construction. Other influences are already beginning to have an impact on the demand for reconditioned goods, RCBPs and recycled materials, even when this may impose some additional costs. These include:

- legislation, especially related to the responsibility for and cost of waste disposal;
- planning policy at local authority level, which influences planning conditions, for example environmental design guidance, waste targets and the Demolition Protocol in the UK;
- achieving credits in environmental assessment tools (e.g. BREEAM in UK, LEED in US);
- the wishes of the client, perhaps encouraged by members of the project team.

In conclusion

There is already a market for reclaimed goods and building materials and products made containing a proportion of recycled materials. However, it is not as mature as the market for new goods and materials. Finding out about the goods available can be difficult, though the internet has made this much easier. Supply can be unreliable and costs unpredictable. The market will become more reliable as it grows. It is best stimulated, perhaps, by examples of what can be done, such as those presented in the next chapter.

2 Case Studies of Reuse and Recycling

Swedish student accommodation made from reclaimed materials

The Swedish town of Linköping was the location of two projects where redundant buildings were used as sources of materials and components to build new student accommodation. These included fitted components such as windows and sinks, and reinforced concrete structural elements and brickwork.

The Udden project

When Linköping had a need for new student accommodation in 1997, the town's largest housing association decided to construct it using materials from two empty 1960s apartment buildings in a nearby town. These had been scheduled for demolition because of severe economic decline in the area. The buildings were constructed mainly from cast in situ reinforced concrete, which was cut into manageable pieces with a diamond saw. Materials taken from about 50 large apartments were used to create 22 smaller student apartments. The major elements reused were:

- 73 concrete wall elements;
- 41 concrete floor beams;
- 30m^2 of concrete foundations;
- 220m^2 clay brickwork;
- 236m^3 mineral wool insulation;
- 636m^2 mineral board insulation;
- 600m^2 woodblock flooring;
- 63m^2 radiators;
- 45 doors;
- 89 windows;
- 26 window sills;
- 78 wardrobes;
- 92 kitchen cupboards;
- 12 kitchen sinks;
- 39 taps;
- 46 toilet basins.

A key contributor to the success of the project was that the same main contractor, Vallonbygden AB, undertook the deconstruction of the old buildings and construction of the new apartments. Nevertheless, the high labour costs meant that the cost of the structural frame was over 75 per cent more expensive than the same frame constructed with virgin materials.

An analysis of the project confirmed that the environmental impact of the reuse option undertaken was less than it would have been using conventional construction techniques and materials, though the reuse option would have resulted in more emissions of nitrous oxide if the materials had been transported more than 140km by lorry.

The Nya Udden project

In 2001 Linköping had a need for yet more student accommodation, and materials were taken from a number of 1970s buildings in a nearby town that were being transformed by having their top storeys removed. The buildings were constructed using pre-cast elements and so could be deconstructed without having to cut through the reinforced concrete elements. Over 400 pre-cast concrete elements were salvaged and reused to construct a new building with 54 small apartments in the Nya Udden project:

- 138 concrete partition walls (524 tonnes);
- 72 concrete outer-wall elements (208 tonnes);
- 224 concrete beams (684 tonnes);
- 8 concrete staircases (16 tonnes);
- 34 windows;
- 100 window sills;

- 16 lengths of steel banisters.

The process proved more costly than anticipated, and the remaining 400 new apartments were built using conventional concrete construction. One contributing factor was probably that different contractors were employed to undertake the deconstruction and assembly of the concrete elements. This led to delays, lack of overall coordination and difficulties with storage of the goods between deconstruction and reuse – the firm deconstructing the elements had little incentive to handle them carefully and ensure they were not damaged.

Both projects cost between 10 and 15 per cent more than similar buildings made using conventional techniques and materials. The shortfall was made up by government grants available for developing new environmentally responsible construction methods. The contractors were confident that with the experience they gained and the increased savings that could be achieved when undertaking such work on a larger scale, they would be able to reduce costs on future projects to make them cost no more than new construction (te Dorsthorst and Kowalczyk, 2002; Eklund et al, 2003).

Lessons learned

Despite less than total success, many useful lessons were learned from both projects. Although none of the contractors had previous experience in building with these types of reclaimed materials, they all found there were no significant structural or other technical difficulties with the deconstruction or reassembly of the concrete materials. The main lesson learned was that a greater proportion of reuse could have been achieved if the sizes of the apartments in the deconstructed and new buildings had been kept the same. In both projects it was found that the new buildings fell short on acoustic performance for sound transmission between adjacent rooms and this was overcome in one case by adding a layer of concrete render and in another case by using gypsum plasterboard. A layer of insulation was also added to the external walls to meet the current building regulations, which were more stringent than when the original building was constructed.

Location: Linköping, Sweden
Date: 1999–2000
Client: Linköping University and Stangastaden Housing Company
Building Engineer: Sundbaums, Linköping
Sponsorship: The National Board of Housing, Sweden
Ref.: Roth and Eklund, 2001
Website: www.stangastaden.se/CM/Templates/Articles/general.aspx?cmguid=eab6ce81-908c-456e-9235-c403b48d2982

BedZED, London, UK

The Beddington Zero Energy Development (BedZED) is an ultra-green residential and office development in the construction of which numerous reclaimed building elements and materials were used (Figure 2.1).

At BedZED, a variety of different methods were implemented to incorporate reclaimed or recycled materials into the construction. A total of 3404 tonnes of alternative materials, including 1862 tonnes of reclaimed on-site sub-grade fill, was sourced and utilized – around 15 per cent by weight of the total materials used in the project. The various reclaimed goods and recycled materials are summarized in Table 2.1.

All the successful measures implemented at BedZED to source and utilize reclaimed or recycled materials resulted in cost savings for the client or the contractor. Three of the measures – the use of reclaimed timber for external studwork, reclaimed doors and reclaimed paving slabs – proved to be too difficult to achieve fully within the constraints of time and budget.

The construction programme at BedZED had to allow a degree of flexibility as finding suitable materials was sometimes difficult and sometimes required a little luck.

Figure 2.1 Beddington Zero Energy Development, South London

Source: Bill Addis

Reclaimed structural steel

An easy, yet cost-neutral solution to utilize reclaimed structural steel on the BedZED development proved successful. Altogether, 98 tonnes of structural steel (95 per cent) was reclaimed from demolition sites within relatively close proximity to the development.

The structural design engineers provided specifications for a range of sections that could be used in each situation. Once appropriate steel sections had been sourced, they were quality checked before purchasing. This included assessing or identifying the date of manufacture, the condition (e.g. any rust or scaling), the number of existing connections and whether bolted or welded, and the suitability for fabrication. The steel was then sandblasted, fabricated and painted. All reclaimed steel products required an extra pass through the sandblaster and treatment with a zinc coating.

Reclaimed steel could not be utilized for curved sections as the local steel contractor refused to pass reclaimed steel through the bending machine. Also, some steel sections were not found as there are currently few reclamation yards with suitable quantities of good structural steel in all sizes.

Table 2.1 Summary of reclaimed and recycled materials used

Material	Off-the-shelf product	Achieved on BedZED	Difficulty	Cost implications
Reclaimed steel	No	Yes	Fairly easy	Saving
Reclaimed timber for internal studwork	No	Yes	Fairly easy	Saving
Reclaimed timber for external studwork	No	No (small quantity)	Difficult	Cost premium
Reclaimed floorboards	No	Yes	Easy	Saving
Reclaimed bollards	Yes	Yes	Easy	Saving
Recycled aggregate	Yes	Yes	Fairly easy	Saving
Recycled crushed green glass sand	Yes	Yes	Easy	Saving
Reclaimed doors	No	No	Difficult	Cheaper than equal quality but more expensive than DIY centre
Reclaimed paving slabs	No	No	Difficult	Neutral (with storage space)
Reclaimed shuttering ply	No	Yes	Easy	Saving
Reused sub-grade fill	–	Yes	Easy	Saving

Note: DIY = 'do it yourself'.

Figure 2.2 Reclaimed steel columns and beams at BedZED

Source: Bill Addis

The construction manager purchased the reclaimed steel on behalf of the client. Although this was a risk to the client, the risk associated with the structural integrity of the steel effectively rested with the structural engineer. The cost of the reclaimed steel was 4 per cent lower than that of new steel (Figure 2.2).

Reclaimed timber

Reclaimed timber for internal studwork

The sourcing of reclaimed timber for the internal studwork on the BedZED scheme was quite easy to accomplish in comparison with other materials. Timber for the studwork was sourced via a large local reclamation yard that also prepared the timber and delivered it directly to site ready for use. The timber supplier provided a guarantee that all timber would be nail-free. The supplier was also responsible for holding the risk associated with the quality of the timber.

Quality was not a major issue with the internal studwork as it is neither structural nor exposed to view or weather. After it has been installed the timber is also no longer visible. The contractors on site had no issues with using the reclaimed timber and have considered using it again on future projects.

The project required 54,000m of 50 × 100mm and 75 × 100mm reclaimed timber for the internal studwork and, due to the size of the order, a competitive price was arranged with the reclamation yard. When accounting for the costs of sourcing and purchasing reclaimed timber, the result of this process was that the project was able to save UK£3350 over the cost of purchasing new softwood.

Reclaimed timber for external studwork

The use of reclaimed timber for the external studwork to BedZED was limited to a small proportion due to time and budget constraints. The remainder was finished in new UK-grown Forestry Stewardship Council (FSC) certified timber.

As the timber for the external studwork had to perform structurally, it was required to be classified as C16 grade. The reclaimed timber needed to be visually stress-graded by a specialist contractor at the reclamation site. This classification proved difficult due to the various ages and mixture of species.

Under National House-Building Council (NHBC) guidelines, some of the species of reclaimed timber also required treatment before being installed. For this project it proved cheaper to treat all the timber rather than separate the species.

The reclaimed timber used for the project was purchased at a lower rate than new timber. However, by the time the additional costs associated with stress-grading and treatment were added, reclaimed timber proved to be the more expensive option.

Reclaimed floorboards

Reclaimed floorboards were used throughout the BedZED development. These boards known as 'onion timber' were bought directly through the joinery contractor and had previously been used as spacers between crates on cargo ships. Using this reclaimed timber was cheaper than using new.

Bollards made from reclaimed railway sleepers

All the bollards at BedZED were made to order from old railway sleepers. The installed cost was cheaper than using new timber.

Reclaimed doors

On the BedZED project, the use of reclaimed doors proved to be a difficult process and did not yield a financial advantage. A total of 476 internal doors were required at BedZED, of which 350 could be reclaimed. To source such a large quantity of doors, it was necessary to establish a supply chain, as no single supplier could fulfil an order of this size. A number of reclamation yards were approached and specifications for the doors provided. It was also agreed to purchase a large number of doors from each yard to achieve some economy of scale.

Once an appropriate door was found, the yard would then arrange for it to be stripped. After the stripping process was completed, the doors were then delivered to the joinery contractor for finishing. The completed the cost of each door, excluding ironmongery, averaged at £67 per door via this supply chain.

Once the supply chain was established, doors were supplied to the BedZED site at a rate of about 20 per week. Quality control proved to be difficult, and many sub-standard doors were getting passed through the chain. In one particular batch of doors delivered to site, 50 per cent were rejected. This poor quality control and the additional staff time to keep the process running smoothly led to the scheme being aborted in favour of lower quality cheaper new doors purchased from a do-it-yourself (DIY) store at £25 each.

This exercise demonstrated that for reclaimed doors to be viable in large-scale building projects, it is important to try to deal with a single supplier that can handle the bulk order. This ensures that the quality of each door is consistent and in line with the specification, and a significant economy of scale can be achieved. The costs associated with reclaiming doors are always likely to be higher than using a lesser quality new product, although the higher quality of the finished reclaimed door can be more attractive to the end user.

General issues

The BedZED project team discovered a number of general issues regarding the use of reclaimed timber products in different building applications. The principal factors that contributed to the success of using reclaimed timber were:

- flexibility in the construction programme to allow long lead times for sourcing the timber;
- storage space on-site to enable bulk orders of materials once it became available;
- providing early design information and cutting schedules.

The use of reclaimed timber for the external studwork was cost prohibitive due to the expense associated with stress-grading and treatment. Costs could be reduced on these items through good organization and it is recommended that large batches of similar species of timber are sourced by the reclamation yards for treatment at one time. By batching the timber in this way, the number of visits required by a qualified stress-grader can also be reduced.

Further reductions in costs could be achieved by designing the timber in common section sizes, allowing the size of orders to be increased. It would also help to be flexible concerning the lengths of timber sections ordered in order to reduce cutting at the reclamation yard.

Reclaimed and recycled minerals

Paving slabs

Some 1800m^2 (around 270 tonnes) of paving slabs were needed for the hard landscaping works at the BedZED development. A source of reclaimed slabs was identified in the local authorities who run pro grammes to replace old or damaged paving. Undamaged slabs, which are usually discarded along with the damaged ones during this process, can be salvaged and stockpiled for reuse. The cheapest quotation for a batch of 490 reclaimed slabs was found to be £1.79 per slab, including packaging and delivery, which compared well with the cost of a new slab at £2 each.

However, research early in the project established that there was no viable way to utilize reclaimed paving. The costs involved with handling and storing paving slabs are very high. In the case of BedZED, limited storage space on-site required that the slabs be stored elsewhere. This not only added a storage cost to each slab but also a cost for double-handling, which made the reclaimed product uneconomical to use. There would also have been a long lead time for their procurement.

Reused sub-grade fill

During the earthworks at the BedZED site some 1860 tonnes of gravel were excavated from beneath the topsoil layer. This gravel was reused as sub-grade for the roads around the scheme.

Recycled aggregate

Recycled crushed aggregate was used on the BedZED scheme to replace virgin limestone aggregate in the construction of the road sub-base. This product, made from crushed concrete, was supplied in the same manner and replaced the same quantity of the conventional limestone.

The earthworks contractor for the scheme already had experience with this recycled product from previous contracts and had no difficulties in finding a supplier for the 980 tonnes required or with the placement of the aggregate. A significant saving of £3.50 per tonne, a total of £3430, was achieved through the use of the recycled product.

Recycled crushed green glass sand

The BedZED scheme utilized 279 tonnes of recycled crushed green glass. This recycled product was used to replace the same quantity of virgin bedding sand required for laying paving slabs. Risk assessments were undertaken as the ground glass is similar to sand but may be sharper to the touch. The problem was avoided by the wearing of gloves and no safety issues occurred.

The recycled sand was easily obtainable from a local aggregate supplier and was approximately £2 per tonne cheaper than the virgin material. With no additional work for the contractor this gave a total cost saving for the material of £558 (approximately 15 per cent).

Lessons learned

While very few technical barriers were encountered, considerable effort was needed to deal with the unconventional processes of procuring the materials for the building and ensuring that sufficient quantities of suitable materials were available when needed.

Location: Sutton, South London, UK
Date: 2002
Client: Peabody Trust
Architect: Bill Dunster with BioRegional Development Group
Services engineer: Ove Arup and Partners
Structural engineer: Ellis & Moore
Ref.: Lazarus, 2002
Website: http://www.zedfactory.com

The C.K. Choi Building, University of British Columbia, Canada

When preparing the project brief for the new C.K. Choi building to house the Institute of Asian Research (see Figure 2.3), the University of British Columbia decided it should be a 'demonstration green building' that would 'set new standards for sustainable design, construction and operation'. The materials of construction were one of seven categories of green design, and targets were set that 50 per cent by weight of the construction materials should be reclaimed or recycled, and that 50 per cent of the materials should be recyclable.

Figure 2.3 Choi building: Exterior, showing reclaimed brickwork

Source: Johnny Winter, Edward Cullinen Architects

It was quickly realized that to meet this target, it would be important to incorporate a significant amount of reused material in the large-scale structural and surface components of the building. The architects established a network of links with local demolition contractors and regularly visited buildings scheduled for demolition in search of potentially useful materials.

Once located, research was then required to determine their suitability and acceptability within existing building codes, testing of materials, long-term durability, and technical aspects of detailing with older or alternative products. The design process needed to remain flexible to accommodate the reused materials as they became available. The actual sizes, quantities and required orientation were often not known until the materials were sourced and purchased. The design process therefore tended to resemble that of a building renovation.

For each element of the building construction a range of reuse/recycling options was drawn up and their viability tested. Reasons why some options were ruled out for the Choi building (but could be viable in other projects) included:

- unacceptable to the university building department, for example crushed glass for fill, carpet made from recycled polyethylene terephthalate (PET);
- unfamiliar to the local building industry, for example use of crushed concrete as fill;
- information about recycled content not available, for example aluminium window frames;
- product not approved by contractors' association, hence not acceptable to university building department, for example roofing system using recycled plastic;
- unacceptable performance, for example odour from carpet underlay made from recycled car tyres;
- new material used in preference to reclaimed to ensure warranty, for example gravel ballast for ethylene propylene diene monomer (EPDM) roofing.

Different issues arose for the reclaimed or recycled goods and materials used and various different approaches were needed to achieve success.

Figure 2.4 Choi building: Reclaimed structural timber

Source: Johnny Winter

Sub-grade material

Fill was sourced from excavation work being undertaken on another construction project on the university campus. The geotechnical engineer worked with the contractor to determine a method of excavation that led to minimal disturbance and fill requirements.

Structural system

The load-bearing structure for part of the building was made from reclaimed timber posts and beams (Figures 2.4 and 2.5). Over 65 per cent of the timber elements came from a building being demolished

Figure 2.5 Choi building: Detail of structural timber showing evidence of its previous life

Source: Johnny Winter

elsewhere on the campus. Prior to the structural design, a sample of the 75-year-old timber was assessed and graded for strength and stiffness. Prior to erection all the timbers were assessed and regraded. Initially, the timber graders, who normally grade new timber, rejected all pieces with prominent visual evidence of previous use. The structural engineer got involved and, as he worked closely with the grader, over 90 per cent of the salvaged timber was passed as suitable for structural use. Initial resistance from the local building regulatory authority also had to be overcome. The timber had to be purchased early in the design stage to ensure its availability to the contractor when required, and the initial design had to be revised to accommodate the precise dimensions of the timber pieces purchased. Structural steel for timber connections, concrete reinforcing, steel decks and seismic bracing had a recycled content of at least 75 per cent and no significant problems were encountered during its use.

Building envelope

The majority of the brick walls were made from reclaimed red bricks from a variety of sources (Figure 2.3). These were purchased early in the design phase to ensure their availability when needed, and a sample of the bricks was tested for strength and durability. The strength of the brickwork was assessed assuming only a chemical bond between mortar and bricks. The additional mechanical bond provided by brick cavities (frogs) was not available since not all the reclaimed bricks had cavities.

Various interior fittings

Virtually all interior fixtures and fittings were assessed for the possibility of using reclaimed components or recycled-content building projects. After availability and cost ruled out many options, a significant number of reclaimed/recycled items were incorporated:

- Partitions were made using recycled-content gypsum wallboard, which was found to be cheaper than the virgin product. The facing was made from 100 per cent recycled newsprint. The recycled content of the core, comprising 18 per cent recycled gypsum and 37 per cent recycled paper, could have been higher if a fire rating had not been required.
- Wall insulation was made with 100 per cent recycled cellulose fibre treated with a borate additive as a fire retardant and mould inhibitor, and with other chemicals to inhibit attack by rodents and insects.
- All the wooden doors and door frames were reclaimed from an office building that was being converted to residential use. Of the steel doors and frames, 90 per cent were also reclaimed. These were purchased early in the design stage to ensure their availability and to enable the detailed design to accommodate the various sizes.
- The aluminium supports for the atrium and staircase balustrades were salvaged from a golf clubhouse that was being demolished near the university. These had to be modified a little on site and new glazing was required to suit the modified sections. New base plates were needed to cope with the higher design loadings in the current building code.
- In the kitchen and bathroom areas of the building all sinks, paper-towel dispensers, garbage receptacles and the partitions between toilet cubicles were reclaimed.
- The carpet underlay was a fibre underlay made from 100 per cent pre-consumer recycled fibre from the textile industry.
- The wall tiles in the bathroom and kitchen areas were made using 70 per cent recycled glass from the automotive industry.
- Approximately 40 per cent of the electrical conduit used was reclaimed. It had to be stored in dry conditions to prevent corrosion and was brushed internally prior to use.

Although no detailed assessments were performed, the design team is confident (based on their experience and knowledge of construction) that the reused and recycled content of the Choi Building exceeded the target of 50 per cent. The attention given to reuse and recycling was extended to the waste produced from the construction of the building. Despite initial reluctance on the part of the contractor, the implementation of an effective waste management plan resulted in only 5 per cent being sent to landfill.

Lessons learned

The main value of undertaking this important precedent project lay in:

- promoting awareness of and dialogue concerning the value of using reclaimed goods and recycled-content products;
- demonstrating that reclaimed and recycled goods and materials need not be inferior to virgin equivalents and can be used in non-domestic building projects;
- demonstrating to suppliers of building materials that building design professionals are changing their expectations regarding reclamation and recycling.

The main lessons learned in undertaking the project were these:

- Commitment from the client is paramount in ensuring that the reclamation options are fully investigated and, when appropriate, made to work.
- The traditional design process and responsibilities of an architect do not address management of reused and recycled building materials. Additional time is required to source, evaluate, negotiate purchase and storage agreements, and then incorporate these materials into a building design. There is also the potential of additional liability.
- It was necessary to buy reclaimed goods as soon as possible in order to ensure their availability when needed by the contractor and to ensure that the design could reflect the products that would actually be used. This meant spending money earlier than would normally be the case.
- The decision to approach a project in this manner requires a partnership between consultants and the developer, with each recognizing both the benefits and difficulties. One possibility would be to introduce an additional team member who works closely with the architect and takes on the responsibility for sourcing, testing and purchasing the reclaimed materials and goods and arranges for their storage and delivery to site – perhaps an employee of the contracting firm.

Location: University of British Columbia, Canada
Date: 1996
Client: University of British Columbia
Architect: Matsuzaki Wright Architects Inc.
Structural engineer: Read Jones Christoffersen Ltd
Websites: www.greenbuildingsbc.com ; www.sustain.ubc.ca

The reclaimed vicarage, Birmingham, UK

The wish to design and construct a building entirely from reclaimed components was made achievable when the proposal received UK government funding in 2003. The purpose of the research project was to learn, by doing, the challenges that face designers and builders when working with second-hand components and materials. The first task was to select a suitable building to construct and, of equal importance, someone who wanted one. After much searching a church was found where a new vicarage was needed to replace one that was to be demolished as part of a plan to redevelop the whole site and provide a number of community facilities as well as the new vicarage. A crucial feature of the scheme was that there were no urgent deadlines and the client was able to accept the relatively slow rate of progress that was anticipated.

After an initial scheme for the vicarage had been developed, studies of the various elements and components that would be needed for such a building concluded, almost surprisingly, that nearly every part of the building could be built with reclaimed materials and components (Figure 2.6). The only exceptions were a few goods where human life would be endangered in the event of a failure, for example fire alarms.

The main issue, then, was not *whether* reclaimed goods and materials could be used, but *how* they could be used – how to obtain them, how to refurbish and warranty them, and at what price? It was also soon learned that a vital part of the process was to devise a procedure by which decisions could be made, especially decisions about whether or not to plan to use reclaimed goods for a certain element and whether goods that were located would be acceptable for use in the building.

During the reconstruction phases of the project, the team focused their effort on three main issues: where to source the reclaimed goods, selecting which building elements to construct with reclaimed materials or equipment, and the process for procuring the various reclaimed goods.

Sourcing the reclaimed goods

Investigations into the four main sources of reclaimed goods yielded the following conclusions:

Reuse on site

- This was a cheap and easy option – there were no supply, transport, storage and handling costs.
- Guidance from a reclamation specialist could help identify what materials and goods were worth reclaiming for reuse.
- The demolition and rebuilding programme for the whole vicarage site would determine what materials would be available at each stage. As the new vicarage was to be the first building constructed there would be little scope for reuse on site.

Salvage yards

- Using salvage dealers is more convenient than using other sources, but at a price to cover the dealers' costs in finding, obtaining, transporting, cleaning and storing items for resale (Figure 2.7).
- The architectural salvage sector attracts high-value items. Lower-value items are not being salvaged to the same extent.
- Dealers have different specialisms. Markets vary according to local supply and demand and land prices (for storage). Dealers should be chosen according to items being sought.
- For the vicarage three suitable local yards were identified that stocked items such as cast-iron radiators, kitchen fixtures, doors, flooring, bricks, roofing slates, steel and timber. The yards also contained materials of little relevance to the vicarage including garden ornaments and window frames that would not comply with the current UK building regulations.

Small-scale refurbishments (DIY)

- Much material from domestic DIY and refurbishment activities can be found in skips by the roadside and at local council 'dumps'. While this route can be rewarding when it yields needed items, it cannot be planned in advance and proved not to be a useful source for the vicarage.

Demolition sites and contractors

- Salvage dealers obtain their materials from demolition sites and these would be a useful source for the goods for the vicarage.
- Demolition practices have changed radically in the last decade due to both commercial and health and safety pressures. Processes are now machine-intensive and tend to direct material streams towards recycling rather than reclamation and reuse. Hence, if reclaimed goods were needed the project team would need to influence the demolition process.
- The following issues need to be addressed when sourcing goods and materials from demolition sites (Figure 2.8 and 2.9):
  - locating buildings due for demolition;
  - locating demolition or reclamation contractors who will salvage the required items;
  - identifying and selecting the components and materials required prior to demolition;
  - organizing the transportation, cleaning and storage of reclaimed goods;
  - costs of goods, transportation, refurbishment and storage.

Selecting the elements to construct with reclaimed materials or equipment

The project team held a number of workshops to identify the main considerations for reuse for architectural, structural, mechanical and electrical elements. Typical questions addressed in the workshops included:

- What reconditioning work is required to allow the element to be reused?
- What are the relative costs of supplying the items when new and when second-hand? This should be both a purchase cost only and also an installed cost.
- What barriers are there to reusing the materials or components in terms of conformance to regulations and standards? How can the barriers be overcome?
- What tests are available for certifying materials and components?
- What contamination risks are associated with each material?
- Are there future durability issues? How can the life expectancy of the materials or components be assessed?
- What are the possible uses for various reclaimed materials?
- What are the sources for the materials and components? Is their availability supply or demand led?

The outcome of the workshops was a series of general principles for reuse and a matrix to identify suitable components for reuse. The general principles identified were as follows:

- Items that require specific performance parameters would require testing (structural, water-tightness, thermal insulation etc.); this could be overcome by oversizing/doubling up, and so forth.
- Some easily reusable items are often already reused (bricks, roof tiles, lead etc.). This can lead to their price being inflated; it could be worthwhile targeting less 'popular' materials (e.g. timber, pipework).
- Cheap items that require high labour input to make them reusable are likely to be uneconomic; it could be worthwhile focusing on high-value materials.
- Items that may be difficult to reuse in their original role may have alternative uses.
- It could be worthwhile targeting major refurbishment projects for materials – high profile office makeovers, refurbishment of listed buildings etc.
- The contractor is essential to the success of the project and must be involved in decisions about material and product selection and purchase, and the feasibility of proposed reuse.
- A component matrix should be used to identify the ease with which materials and components could be reused. Cost implications should then be assessed with input from the contractor.

After assessing the potential for using reclaimed components and materials in different building elements, the unique features of the vicarage project were reviewed to identify project-related constraints, including:

- the phasing of demolition, construction and occupation;
- the reuse strategy and its constraints;
- planning authority and other statutory requirements;
- site considerations;
- layout and design of the building, including the need for public and private spaces.

The project team then undertook an analysis of the proposed building element-by-element, component-by-component and material-by-material to identify where it would be most practical to use reclaimed goods. The principal ideas included:

- a large conservatory wall made of reclaimed windows;
- reclaimed timber cladding;
- heavy rubble or masonry walls;
- reclaimed plasterboard and timber panelling for interior walls;
- foundations constructed from compacted recovered crush material, such as inert demolition waste and recycled crushed concrete;

Figure 2.6 The 'reclaimed vicarage', Birmingham, UK: Architect's model

Source: Buro Happold

Figure 2.7 School desktops will be used to form the walls of the vicarage

Source: Buro Happold

Figure 2.8 Suitable windows have been identified in a school to be demolished

Source: Buro Happold

Figure 2.9 Cupboards will be reclaimed from school to be demolished

Source: Buro Happold

- rammed earth inner walls, constructed from soil excavated onsite during construction;
- reclaimed timber joists;
- floorboards constructed of timber or chipboard salvaged from discarded office furniture;
- a reconditioned domestic gas-fired boiler;
- reconditioned conventional steel panel radiators;
- reclaimed gas pipework, pressure-tested to in excess of normal working pressure;
- reclaimed copper pipework for internal cold water and reclaimed medium-density polyethylene (MDPE) pipework to connect the house to the local water main;
- insulation reclaimed from copper pipework for hot water distribution;
- reclaimed rainwater guttering and downpipes;
- reclaimed extract fans for kitchen and toilet;
- reclaimed electrical components including cabling, light fittings, light switches, power sockets, earth wiring, electricity meter and security system.

Following this exercise, a shorter list was drawn up with those elements/items for which the reuse option would be taken further. A note was made of the potential advantages and disadvantages of each option. It was accepted that some of these items would fall off the list if difficulties arose later in the procurement process.

Procuring the reclaimed goods

In order that the decision process for procuring each reclaimed item for the building could be planned and carried out by the entire project team (Table 2.2), a seven-step procedure was developed to cover the various stages up to completion of the building:

1. identification of requirement;
2. decide selection basis for the item;
3. try to source;
4. to buy or not to buy?
5. storage and testing;
6. check/amend design;
7. after installation.

Each step contained a number of decisions to be taken, the outcomes of which were recorded.

Step 1: Identification of requirement

Specify the item needed or, preferably, the performance it needs to meet; for example timber reclaimed from groynes, or long elements to form the sides of the walkways on the ramp.

Specify the quantity needed, and any other absolute requirements; for example 60 linear metres, 1 metre high, strong enough to carry self-weight over 2.5 metres.

Step 2: Decide selection basis for the item

Agree criteria for the item, for example:

- embodied energy (including transport);
- toxicity;
- cost of purchase;
- cost of labour to prepare/test;
- cost of labour to install;
- cost of operations (including maintenance and refurbishment);
- further performance requirements (e.g. aesthetics).

Table 2.2 Sample table for assigning tasks

	Architect	Engineer	Builder	Client
Demolition contractors				
Salvage firm				
Telephone likely sources				
Contact planners				
Contact demolition contractors				

Table 2.3 Factors influencing choice of purchase

	Threshold level	Found material
1 Cost to buy		
2 Cost to install		
3 Transport miles		
4 Toxicity		
5 Fitness for purpose		

Decision made:	Yes/no	Date

Set threshold levels for these that would be acceptable (Table 2.3), for example:

- total cost below UK£XX;
- embodied energy saved (energy saved less transport impact) XX;
- toxicity must be below threshold XX;
- fit for purpose XX.

Step 3: Try to source

Assign members of the project team to pursue the different possible sources for each reclaimed component/material identified for use.

Step 4: To buy or not to buy?

Depending on the importance of the item, a decision will be needed as to who can make the choice as to whether or not to buy. This must tie in with design responsibilities and the team member who is trying to source them. The team must be kept informed, but not slow down the process. Ultimately the building contractor must be involved in all decisions to buy to ensure the suitability for use.

If one of the sources produces a material that meets the thresholds then it can be bought. If the costs are considered too high, then the purchase will need to be reconsidered, and perhaps more people need to be involved. Where a number of possible sources arise then a choice must be made between them (Table 2.4).

Finally, a decision must be made as to who will place the order (expect to move towards this being the builder).

Step 5: Storage and testing

Decide where to store the components/materials and where to refurbish and test components and items of equipment:

- at their original source;
- at an intermediate storage depot;
- on the vicarage construction site.

Step 6: Check/amend design

Each reclaimed item and material may have an impact on the design, owing to its actual dimensions, fixings, condition, constituent materials, and so on. Each member of the design team will need to check that this does not have an adverse effect on the design and/or adjust the design to suit the items found.

Table 2.4 Purchase decision-making

	Architect	Engineer	Builder	Client
Component 1	Y/N	Y/N	Y/N	Y/N
Component 2	Y/N	Y/N	Y/N	Y/N

Step 7: After installation

Compile a report for each reclaimed item or material on:

- final total costs;
- the effectiveness of the reuse of this element.

Lessons learned in pre-construction phase

The main lesson learned during the design phase was that designing a building to be constructed with reclaimed components is utterly different from the normal design process, in that suppliers need to be identified before design is completed.

Work on the construction of the reclaimed vicarage started in September 2005 and is scheduled for completion around Easter 2006. As the construction phase approaches it is clear that it will nearly resemble a 'normal' project as a result of the thorough procedure used in the pre-construction phases.

Location: Birmingham, UK
Client: All Saints Church, Kings Heath, Birmingham
Date: 2003–06
Project leader and design engineers: Buro Happold Consulting Engineers
Architect: Cottrell and Vermeulen Architects
Reclamation consultant: Salvo
Main contractor: Skanska
Funding: UK Department of Trade and Industry

Using recycled-content building products in the US

Southern California Gas Company Energy Resource Centre

The design of the new Energy Resource Centre (ERC) building was a refurbishment of an existing office building constructed in 1957. One of the primary goals of the Southern California Gas Company was to use as many reclaimed or recycled materials in the construction of the ERC as possible. About a third of the existing building was demolished to allow construction of the new 4140m^2 building comprising office space, meeting rooms and an exhibition hall. The ERC incorporates many green-design features including not only low energy use but also the use of a number of reclaimed goods and recycled-content building products.

The main driver for these initiatives was the company's desire to make a good impression with its customers and shareholders, demonstrating its commitment to the conservation of natural resources in the design and construction of their new building.

The previous company office building was dismantled to preserve materials that could be reused in the new building. About 60 per cent of the 550 tonnes of demolition materials reclaimed from the old office building was either reused in the new building or recycled in other ways. In addition to the reuse of materials from the old building, 80 per cent of all construction materials used in the new ERC building were reclaimed, contained recycled materials or were made from renewable resources.

Among the reclaimed or recycled materials used in the ERC:

- An entire staircase structure was reclaimed from the set of a film studio. It was modified in minor ways to ensure it satisfied all the relevant building codes.
- The wooden floorboards to the reception area were reclaimed from a condemned warehouse in San Francisco.
- The carpeting is made from 50 per cent recycled-content carpet that can be recycled again at the end of its life.
- A decorative wall in the reception area is constructed using 100 per cent recycled aluminium from the aircraft industry.
- The concrete mix for the entry walkway to the building was coloured by adding PVC chippings made from the offcuts of gas pipes.
- A decorative countertop in the reception area was made from 100 per cent glass recycled from windows.

- Exhibition panels in the display area in the main hall are made of 100 per cent recycled paper.
- The acoustic ceiling tiles in seminar rooms contain approximately 22 per cent post-consumer waste.
- The partitions used in the toilets are made from 50–90 per cent recycled plastic.
- The filler for concrete expansion joints in the building was made using 100 per cent recycled newsprint.
- The floor tiles for the complex contain 70 per cent post-consumer glass.

All the reclaimed or recycled materials incorporated in the building were assessed for their financial feasibility before they were adopted. The client has estimated a saving of US$3,000,000 by using the environmentally friendly products and using materials reclaimed from the previous office building.

Location: Downey, California
Date: 1995
Client: Southern California Gas Company
Architect: Wolff Lang Christopher Architects
Contractor: Turner Construction Company
Websites: www. nrc-recycle.org/brba/casestudies/energy.htm
www.ciwmb.ca.gov/publications/greenbuilding/42296043.doc
www.eere.energy.gov/buildings/highperformance/case_studies/overview.cfm

Duracell world headquarters, Connecticut

The new headquarters for Duracell Corporation in Connecticut comprises three levels of office space, a conference centre and research laboratories. As part of its commitment to environmental sustainability, the client made the commitment to achieve the gold rating using the environmental assessment tool, LEED. Several LEED credits were gained by ensuring that as many reclaimed fittings and recycled materials as possible would be used. At the beginning of the project a target was set for a minimum of 50 per cent of all construction materials to have a recycled content (of unspecified amount). Though very little reclaimed components or materials were used, the target was met using the following products:

- foundation drainage: 50 per cent recycled PVC;
- acoustic wall panels: 50 per cent recycled vinyl mounting track and 65 per cent recycled slag tackable boards;
- foam insulation: 3 per cent recycled plastic foam;
- asphalt paving: minimum 15 per cent recycled asphalt;
- cast-in-situ concrete: 50 per cent blast furnace by-product;
- pre-cast concrete: 30 per cent recycled blast furnace by-product;
- flush wood doors: 50 per cent recycled hardwood;
- ceiling tiles: 18–90 per cent recycled newsprint, wood fibres and slag wool;
- fireproofing: 65 per cent recycled paper;
- fibreglass insulation: 3–5 per cent recycled glass;
- fire protection: 100 per cent recycled glass;
- interior stonework: 100 per cent recycled glass tiles;
- tiles: 70 per cent recycled glass.

Location: Bethel, Connecticut, US
Date: 1995
Client: Duracell
Architect: Herbert S. Newman Associates
Contractor: Turner Construction Company
Websites: www.hsnparch.com/newmanFrameX.html
www.brba-epp.org/brba-epp.org/pdfs/Durcell.pdf

Lessons learned

The incorporation of credits for using RCBPs in the LEED green building assessment tool provided a useful stimulus for the project team. The task of procuring RCPBs was made easier by the availability of good product data including certified information about the proportion of recycled materials in the various products. Despite using many RCPBs, their weight was small in comparison to the total weight of materials – mainly steel, concrete and glass – used in the buildings.

BRE Building 16: The 'energy-efficient office of the future', Garston, UK

In the early 1990s the Building Research Establishment, then a government research station, decided to build a new building in Garston in the UK that would demonstrate best practice in environmental design. Although the best-known feature of the building is its low energy use – around half that of an equivalent building designed and constructed in conventional ways – it is no less remarkable for the determined effort that was made to use reclaimed goods and reclaimed and recycled materials (Hobbs and Collins, 1997).

Demolition of the existing building

First, a reclamation audit was undertaken on the building that occupied the site. This determined the items and materials that had potential for reclamation in the reclamation market. Items such as furniture, timber off-cuts, light fittings, venetian blinds, fire-alarm equipment, heaters, dryers, light switches and sockets were removed in the soft strip and either reused on the BRE site or donated to a charity organization for onward supply to local schools and hospitals. Next a number of materials were reclaimed, including slate cladding, roofing sheets, cast-iron rainwater goods and roofing timbers (which were sold to a furniture manufacturer).

It was not possible to reclaim the 20,000 or so bricks as they used modern cement mortar (rather than lime mortar) and separation was not practical. They were crushed and used on site as hardcore. Doors could not be reused in the new building because they did not meet modern fire safety regulations. While the doors were of suitable quality for reuse, insufficient time was available to find buyers and they were sent to a landfill site. Despite this, around 95 per cent by volume of the materials in the existing building were reused or recycled.

Construction of the new building

Reclaimed bricks

Around 80,000 reclaimed bricks were used to clad the new building and supplied from a reclamation dealer less than 100km away. A plan to use bricks reclaimed from a hospital less than 2km from the BRE was frustrated when demolition of this building was delayed. The use of the reclaimed bricks was more costly than the new equivalent. Not only did they cost more, but the architect needed additional design time to cater for the change from metric to imperial sizes and the contractor had to spend additional time laying the reclaimed imperial bricks within a metric grid.

Reclaimed flooring

Some 300 m^2 of wood-block flooring was used in the new building and this was all obtained by the contractor from another of the firm's projects where a substantial quantity of parquet flooring was reclaimed. Although more expensive to lay than new wood-block flooring, a cost saving of around 30 per cent was achieved.

Recycled concrete aggregate

Although useful quantities of crushed aggregate were produced when the previous building was demolished, it was not used to make concrete – it was all used as fill or hardcore material on the site, and too little time was available for crushing between the demolition materials becoming available and the need for new concrete. The crushed concrete for use in the building came from a 12-storey office building being demolished in central London, some 20 kilometres away. It was crushed in two stages to provide aggregate in the 5–20mm range and delivered to a local ready-mix concrete firm. This source provided all the coarse aggregate for 1500cm^3 of concrete specified as a C25 or C35 mix (minimum strength of 25 or 35N/mm^2) and used for foundations, floor slabs, structural columns and waffle-slab floors. All mixes contained 985kg/m^3 of crushed aggregate except those that were placed by concrete pumping, which contained 935kg/m^3. Ground granulated blast-furnace slag (GGBS) was used as cement replacement in both mixes to provide protection against carbonation. A larger than usual number of trial mixes was made to establish the most suitable mix design.

Lessons learned

A number of lessons were learned that helped identify various subjects for further research that the BRE has been undertaking since the completion of Building 16:

- It is important to understand the reclamation industry (better) before embarking on the reuse of goods and materials.
- Only the use of lime-based mortars will allow brickwork to be reclaimed.
- It is important to allow enough time in the construction programme to arrange for disposal of goods and materials for reuse or reclamation, and to acquire reclaimed goods and materials for a new project. A better developed waste-exchange infrastructure will greatly help achieve these aims.

Location: Garston, UK
Date: 1997
Client: Building Research Establishment
Architect: Fielden Clegg
Services engineer: Max Fordham and Partners
Structural engineer: Buro Happold
Reclamation audit: Salvo
Contractor: John Sisk
Refs.: Hobbs and Collins, 1997

The Earth Centre, Sheffield, UK

The Earth Centre, built on the site of two former collieries, was conceived as a visitor attraction and country park to demonstrate a variety of green building ideas. It was decided that the new conference centre and entrance building should be exemplar buildings and, in addition to many other green construction ideas such as low energy use, there was a commitment to use reclaimed components and materials wherever possible. The project team took the decision to concentrate their reuse and recycling efforts in five elements: gabions, concrete, reclaimed timber, insulation and radiators.

Gabions were used both as retaining walls and for some of the walls to the building (Figure 2.10). They were chosen in preference to alternatives for several reasons – they were cheaper than reinforced concrete, more aesthetically suitable than concrete blockwork, and more reliable as a structure material than rammed earth. It had been the original intention to fill the gabions with local sandstone but this idea was rejected in favour of crushed concrete because of the adverse environmental impact of using the natural stone. The recycled crushed concrete was sourced from buildings being demolished less than 20km from the site.

Reclaimed timber, sourced just 3km from the site, was used for the roof structure. However, some difficulties were encountered in making it ready for

Figure 2.10 The Earth Centre, near Sheffield, UK; reclaimed radiators fixed to walls made for gabions filled with recycled concrete

Source: Taylor Woodrow

use in the building. No firm could be found who would take the risk of damaging their lathes to turn and finish the reclaimed telegraph poles – the firms felt there was too great a chance of encountering embedded nails or screws. To resolve the problem a makeshift lathe was constructed, but at the expense of much time and effort. The storage of the reclaimed timber between purchase and its use had to be carefully managed. While it was satisfactory to store the timber outside, this led to a moisture content well above what was acceptable for its use. The timber had to be brought undercover well in advance of when it was to be used to give it time to dry out and reach an acceptable moisture content. A further hurdle that the project team had to overcome was the nature of the reclaimed timber market, which differed significantly from conventional timber merchants. Not only was it was necessary to haggle over prices that always started well above a sensible value, it was also necessary to be able to buy the materials on the spot. It was important to develop a good working relationship with the suppliers who would then be willing to look out for desired goods and even hold material until it was needed.

The concrete-piled foundations and ground beams were made with a cement containing 70 per cent GGBS. This concrete took twice as long to reach its design strength of 35N/mm^2 – 56 days rather than 28 days – but this had been allowed for in the construction programme. A possible problem with using GGBS was identified, though did not in fact affect this project. There are very few GGBS batching plants (which have to be near steel works) and this can increase the cost and environmental impact of transport; also there is no continuous and unlimited supply of GGBS and the project team heard of other projects where GGBS could not be obtained when it was needed. The use of recycled aggregates was investigated but not carried out – the only convenient supply was from a source two hours away and the price quoted was twice that for virgin aggregate.

Insulation made from recycled newsprint was used. This material is pumped into position slightly wet, which activates a natural resin in the paper that hardens and has the effect of fixing the insulation in place.

Reclaimed radiators can be bought in any condition freshly salvaged from a demolition site, costing just UK£30–40 each if bought in bulk, though rising to upwards of £250 if fully refurbished and tested. A purchaser has to choose the level of risk they are prepared to accept and work they are prepared to undertake themselves. The Earth Centre project team located an adequate source of radiators in a hospital due for demolition and which were believed to be in good condition. However, this demolition contract was delayed and, at the last minute, the Earth Centre project team had to find an alternative supplier – a dealer in antique fittings – for the first batch of radiators to keep the project on schedule. The refurbishing of the radiators consisted of pressure-testing them to three atmospheric pressures, flushing them to remove silt, fitting new connections, removing old paint and repainting. In the first batch from the antique dealer, one in ten radiators had to be rejected. Of the remaining radiators salvaged from the hospital a quarter had to be rejected – a statistic that was believed to arise mainly from damage caused in the salvaging operation. This suggests that it would have been better to set up an alternative arrangement whereby the Earth Centre contractor would have undertaken the removal from the hospital of the equipment needed for their project. The flushing of the radiators also proved more problematic than anticipated. Once installed and in operation, more silt that had been left behind after the flushing operation was released and blocked the heat exchangers. The addition of filters solved this problem but a pump had to be installed in what had been a gravity-driven heating system.

Lessons learned

The key lesson learned from this project was that reuse and use of reclaimed materials is a realistic option for many building elements but a significant amount of additional work is required by the project team that would not be encountered when using 'normal' materials and goods. Most essential of all was the resolve to achieve the goal set at the beginning of the project.

Location: Near Doncaster, UK
Date: 1999
Client: The Earth Centre
Architect: Bill Dunster
Services engineer: Ove Arup and Partners
Structural engineer: Mark Lovell Design Engineers
Main contractor: Taylor Woodrow
Website: www.zedfactory.com

Westborough School, Southend, UK

When a new playroom was planned for Westborough Primary School, it was decided to use the opportunity to construct an experimental building with the main aim of exploring the opportunities for using cardboard in building construction (Figure 2.11). The experimental nature of the project attracted funding from the UK government.

The main benefit of using cardboard is that it is a material with a very high recycled content – often over 90 per cent. Furthermore the material can be recycled again once the products have served their useful life. The source of the fibres for the cardboard used in this project was a mixture of post-industrial and post-consumer paper and board.

A secondary aim of the project was to use 90 per cent (by weight) of recycled or easily recyclable materials (not including the concrete used for the floor slab). A level of 50 per cent was achieved, mainly because the quantities of timber components in the building were larger than originally anticipated and virgin timber was used. Table 2.5 illustrates the recycled-content components used.

Figure 2.11 New classroom at Westborough School, constructed largely from cardboard made from recycled paper

Source: Buro Happold

Cardboard as a building material

A new or unconventional material cannot be used in building without considerable work to establish its suitability. The issues described below for elements made from recycled-content cardboard would need to be addressed for any building elements made from unconventional building materials, whether recycled or simply unusual.

Stiffness and strength data

The two main properties that influence the quantity of material used for structural elements are its strength and stiffness. Test data were found that allowed the structural engineers to design the load-bearing elements with confidence, subject to a number of issues (below) that distinguish cardboard from more familiar materials such as steel and timber.

Table 2.5 Recycled materials used at Westborough School

Material	Item	Volume m^3	Weight kg	Recycled content %
Cardboard panels	External walls, roof, some internal partitions	22.5	2260	100
Cardboard tubes	Columns, ventilation ducts	0.5	400	100
Sasmox board	Fireproof facing walls	0.2	240	50
Sundeala board	Acoustic lining panels	0.4	480	100
Rubber flooring	Floor finish for entire building	0.5	675	100
Tectan (recycled fruit juice cartons	Kitchen cupboard doors and splashboards	0.12	200	100

Source: Buro Happold (unpublished data)

Figure 2.12 Interior showing columns made from recycled paper (as cardboard) and kitchen worktop and cupboard doors made from recycled plastic

Source: Buro Happold

Figure 2.13 Ventilation duct in toilet made from recycled paper (as cardboard) and cubicle partitions made from recycled plastic

Source: Buro Happold

Effect of creep

Like timber, paper tubes are susceptible to visco-elastic behaviour or creep, that is, an increasing deflection over time during the application of a fixed load. The tests carried out on tubes have found that creep is negligible when loads are limited to 10 per cent of the compressive strength. This value was used in the design of the load-bearing elements.

Effect of moisture content

The strength of the cardboard tubes varies with the moisture content and this, in turn, varies with the atmospheric relative humidity. To allow for this, the elements were designed to carry 20 per cent greater design load than usual.

The moisture content also affects the material stiffness (Young's Modulus) with the result that the building elements move slightly or 'breathe' as humidity changes. The two main consequences of this are that deflections of the structure will vary from day to day and that, in the long term, the material may break down as the hydrogen bonds holding the paper together suffer fatigue failure. No long-term test results were found that define this behaviour further and so it was decided to deal with the possible problem by applying a PVC or aluminium foil vapour barrier to the surface of the cardboard to limit ingress of moisture into the load-bearing elements.

Effect of temperature

Tests carried out to date show that both stiffness and strength are reduced at elevated temperatures, most likely attributable to a breakdown of glues and binding agents. This confirmed that cardboard used for structural purposes would need to be protected from significant temperature change, as well as fire, by insulation. In view of the experimental nature of the structural material, it was decided to monitor its behaviour carefully and frequently during the construction period, and intermittently throughout the life of the building. In the few months immediately following completion of the building deflections of some elements were detected but considered to be due to the drying out of the glues used and not a danger.

Effect of fire

Like all wood-based products, paper burns unless treated. Like timber, card tends to char rather than play an active part in a fire and the charring affords fire protection to the material beneath. Nevertheless fire protection is needed. The challenge with fire treatments was not to diminish the good environmental performance that card offers by using chemicals with a detrimental impact on the environment. Tests on the performance of fire-protected cardboard elements demonstrated a very good fire performance. At the Westborough School building an over-cladding solution was adopted for most components. The only exposed card was on the ceilings and was treated to restrict flame spread.

Effect of water

As cardboard becomes a pulp when wetted, it was imperative that all necessary steps are taken to prevent water ingress into the cardboard elements. Untreated card is also hygroscopic, meaning that it absorbs moisture from the air. In the Westborough project a three-step approach was adopted to protect the card panels from water:

- The first level of protection was to use water-resistant cardboard. In the manufacture of the cardboard from pulp, additives can be introduced into the pulp mix that render the board water-resistant. This means that the cardboard inherently has a reduced susceptibility to the effects of moisture and these additives can be removed on re-pulping.
- A second level of protection was to apply an external coating to the various elements. The main source of moisture is warm, moist air on the inside generated by the occupants and their activities. A polymer vapour barrier was applied on the inside, and a breathable building paper water barrier applied on the outside face. This minimizes the flow of water vapour into the card and allows it to escape should any collect inside the card. This reflects normal practice for timber-framed buildings.
- Although a coating layer will be waterproof when it is first installed, it is vulnerable to damage. In order to protect the waterproof layers from contact from both sides, on the inside an extra 1mm layer of board was added after the vapour barrier to physically protect it from scratching. The walls were also protected with a pin-board to guard against damage from impact and, especially, drawing pins! The outside of the building was protected by over-cladding using a material as close to cardboard as was possible – a panel product made from wood fibre and cement.

Damage by insects and mould

While it is possible to protect paper products from damage by insects using a boron-based product, as done with recycled-paper insulation, it was not done in this case because it would render the material non-recyclable. The possible problem was avoided by two lines of defence:

- By keeping water out, most insects and moulds are discouraged.
- Insects are excluded by an insect mesh in the ventilated rainscreen openings.

Acoustics

The location and nature of the building at Westborough School meant there were no specific requirements for the acoustic performance of the cardboard panels. The trapped air in the panels helps to reduce sound transmission, though there is little mass to absorb sound energy. Nevertheless, an adequate acoustic performance (38dB reduction) was achieved.

Security/insurance

Given the basic cardboard is relatively easy to cut, particularly compared to brick or stone, there was a potential security issue. This was solved by the use of the fibre-cement over-cladding layer, and adequate insurance cover has been provided.

Lessons learned

Using unconventional materials for building components requires a first-principles approach by the design engineers. While not technically difficult, this requires engineers with the right attitude and time to allocate during design to develop the necessary understanding of the materials – across many disciplines – and how to use them. Similarly, the design team need to find contractors with a similar spirit of adventure who are prepared to undertake some experimentation prior to final construction.

Location: Southend, Essex, UK
Date: 2000–01
Client: Westborough Primary School
Project manager and all engineering design: Buro Happold
Architectural design: Cottrell & Vermeulen Architecture
Manufacturers of paper and board: Paper Marc Ltd
Manufacturers of tubes: Essex Tube Windings Ltd
Manufacturers of panel products: Quinton and Kaines Ltd
Building contractors: C. G. Franklin Ltd
Funding: UK Department of the Environment, Transport and the Regions; Cory Environmental Trust, Southend, UK
Website: www.cardboardschool.co.uk

Canalside West, Huddersfield, UK

The West Mill in Huddersfield was one of many hundreds of similar buildings built in the 19th century to house Yorkshire's successful spinning and weaving industries. Such buildings characterize former industrial towns in Britain and many other countries and today are seen as valuable to our national heritage. The West Mill was acquired in the 1990s by the Polytechnic (now University) of Huddersfield as part of its expansion plans, and the building was identified as suitable to provide a new home to the Department of Mathematics and Computing.

The six-storey building is constructed of external, load-bearing brick walls with cast-iron beams spanning the full width of the building between the external walls and the central line of cast-iron columns. The floors are stone flags resting on a rubble fill supported on brick jack arches spanning some 3m between the parallel cast-iron beams (Figure 2.14). The roof structure comprises wrought-iron trusses at 1.5m centres and was originally covered with slates, but these were replaced in the early 20th century by asbestos sheeting.

Such industrial buildings have a number of characteristics that make them suitable for a range of modern uses:

- They were designed to carry large floor loadings in excess of modern office floor loadings.
- The height of each storey (around 3.15m) is large enough to accommodate modern building services overhead or beneath a raised floor.
- The structural system is highly suitable for introducing new staircases, lifts and building services risers.
- All the load-bearing elements are either visible or easy to inspect with minimum intervention (easier than reinforced concrete).
- None of the construction materials decay significantly while the building is in use.

The approach taken to appraising the structure and assessing its potential for reuse followed the rational methodology described in Chapter 5. A full survey of the building established the structural system and construction materials and the dimensions of all key elements.

Figure 2.14 Interior of the disused mill prior to reuse of structure and refurbishment

Source: Bill Addis

With these data it was possible to assess the load capacity of the floors. The first calculations indicated that the cast-iron beams and jack-arch floor would not quite be able to carry modern floor loading. The main weakness was found to be the cast-iron beams that seemed unable to carry the tensile stresses in the floor (cast iron is relatively weak in tension). It was also found that that the columns on the first floor were loaded nearly up to their capacity. (The survey revealed that the columns intended for the first and second floors were incorrectly installed, respectively, on the second and first floors – the first-floor columns carry greater loads and need to be stronger than the second-floor columns). If matters had rested here, the conclusion would have been that the structure of the building was inadequate and all the internal cast-iron beams and columns and the jack-arch floors would have to be removed and replaced, probably by steel beams and columns with a profiled metal-decking floor.

Common sense told the engineers that it was very unlikely that the structure of the building was inadequate and a second level of investigation was undertaken. Samples of the brickwork from the jack arches were taken and tested in compression. This showed that the strength originally assumed had been conservative. Samples of the cast iron were taken from the beams and columns and tested in compression and tension. Twelve samples were taken that enabled a 95 per cent (i.e. satisfactory) level of confidence in assumptions about the iron in other beams and columns not tested. These tests showed that the properties originally assumed had been conservative. The precise construction of connections between beams and columns was investigated. This revealed that a slightly shorter length (compared to original assumptions) of both beams and columns could be used in the structural calculations. This increased the predicted capacity of both elements.

These revised calculations still indicated that the structure was not strong enough to carry the floor loadings of the new building. Two additional measures were proposed. First, the dead-load on the cast-iron beams could be reduced by removing the rubble fill and stone flags and replacing it with less-dense, lightweight concrete. Second, the top flange of the floor beams could be encased with a cage of steel reinforcement, which, after the lightweight concrete had been placed, would effectively create a deep, composite beam with a top flange of reinforcement and a bottom flange of cast iron (Figure 2.15). In this way, the *entire* cross-section of iron was made to carry the tension stresses in the floor beams, some 30 per cent more than the area of iron carrying tension in the original floor.

These measures were sufficient to enable the existing structure to be reused. Not only was this substantially cheaper than replacing the internal structure with a steel frame, it also preserved the cast-iron and jack-arch floors that contribute so much to the character of the building (Figure 2.17).

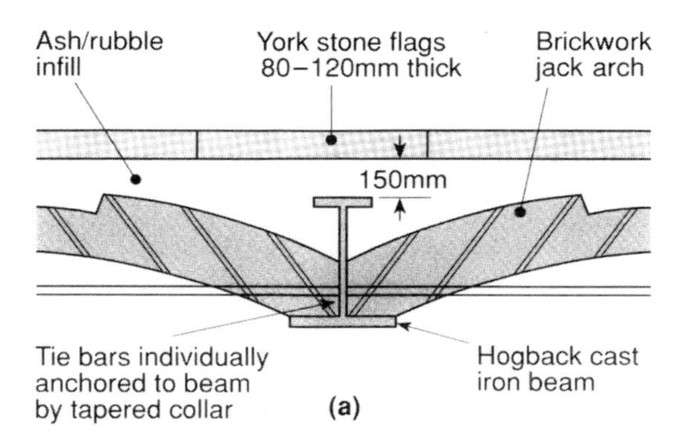

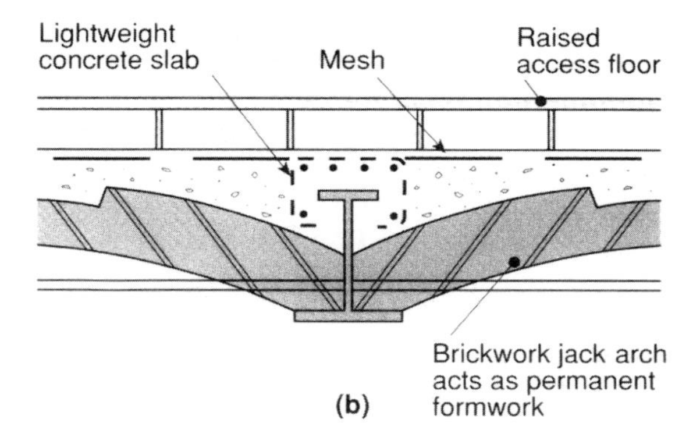

Figure 2.15 Cross section of jack-arch floor showing reduction of dead load and strengthening

Source: Ove Arup and Partners

Figure 2.16 Jack arches were removed in some bays to provide space for new staircases and services risers

Source: Ove Arup and Partners

Figure 2.17 Interior of the refurbished building showing new windows, cleaned jack arches and exposed structure

Source: Bill Addis

Lessons learned

The process of assessing the structure and main fabric of an existing building for reuse is not difficult – it is mainly a matter of attitude. The reuse option should be thoroughly investigated by a competent engineer to ensure that a building is not condemned as 'unsafe' by an engineer who would simply prefer to work on designing a new building rather than reusing an existing one.

Location: Huddersfield, UK
Date: Original building 1865; refurbishment 1993–96
Client: University of Huddersfield
Structural engineer: Ove Arup and Partners
Architect: Peter Wright and Martyn Phelps
Refs: Bussell and Robinson, 1998; Robinson and Marsland, 1996

Reusing structural steelwork

The structural steelwork in buildings constructed since the beginning of the 20th century seldom loses its capacity to carry loads and can usually be disassembled with relative ease. Many steel-frame buildings are now reaching the end of their first lives but are unsuitable for reusing in situ because of low storey heights, the need to introduce modern building services, or simply because they have too few storeys or an unsuitable footprint. There are thus likely to be a growing number of opportunities for salvaging steel sections from buildings being demolished and reusing them in new construction. On a small scale this has been happening for a long time, but undertaking reuse on a larger scale requires the development of a suitable procedure for assessing the potential for reuse and any remedial work that needs to be undertaken.

The UK firm of structural engineers, Ellis & Moore, working in collaboration with BioRegional Reclaimed, gained experience in using salvaged steel sections in the BedZed project. Building on this work they have developed a procedure for recertifying structural elements made of steel.

Environmental impact of steel production

In recent years with more awareness about global warming and the effects of CO_2 emissions, industrial processes that consume large quantities of energy need to be assessed to try to cut carbon emissions and save resources for the future.

Steel producers have been very active in improving the efficiency of their production plants. They already recycle large quantities of existing steel into new steel. By recycling scrap, they reduce the energy consumed and the associated environmental impacts of extracting virgin materials. However, remelting the old steel still requires large amounts of energy so that steel made with a proportion of recycled material has an embodied energy more than half that of new steel. A better alternative is to use reclaimed steel sections where possible rather than melting it down. A comparison of the environmental profile of reclaimed steel compared to new steel clearly shows the benefits (Figure 2.18).

To reduce energy costs as much as possible, it is desirable to transport the reclaimed steel sections directly to site to be fabricated and erected, thereby reducing both the cost and environmental impact of transporting steel. The viability of this approach would depend on the quantities of steel involved as well as the opportunities for providing the necessary workshop facilities on site. Currently fabricators are reluctant to consider this option as they need to justify their plant and overheads, including a labour force that is not mobile.

Assessment procedure

A procedure has been developed that enables the project team to assess the potential for reuse and to achieve this goal by following a series of pragmatic steps. It should be noted, however, that simply following the procedure is not enough. It must be undertaken by a team committed to achieving the intended outcome, from design through to the delivery on site. It also requires a client willing to support the idea of reuse.

In order that second-hand steel be successfully

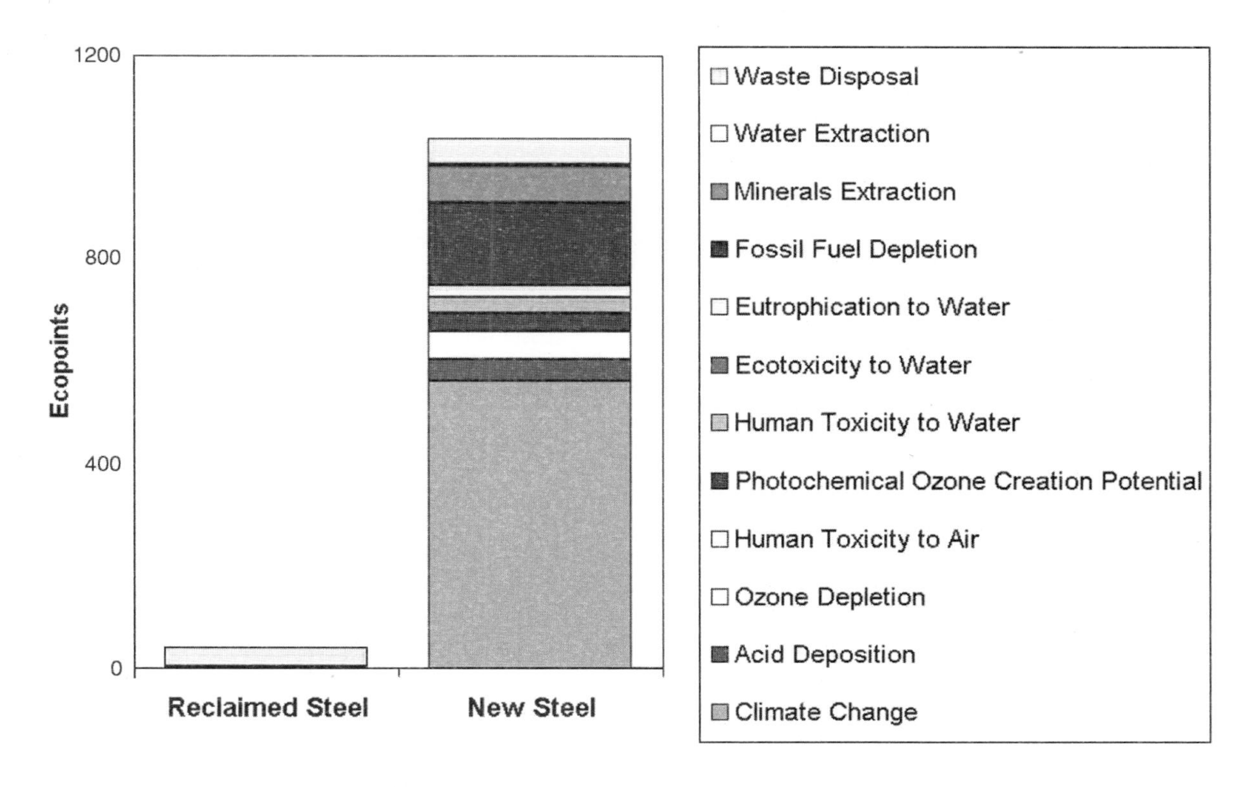

Figure 2.18 Environmental profile of reclaimed steel versus new steel

Source: BioRegional Reclaimed

reused there needs to be a simple process whereby the material is sourced, checked and fabricated for installation in either new-build projects or refurbishments. The steel needs to be equivalent to new steel with the same life expectancy. The client and project team need to be confident that the steel will perform as new steel, in a similar manner to normal refurbishment projects. These issues need to be resolved early to avoid problems with collateral warranties. The procedure also assumes that the main contractor for the project supports the use of reclaimed steel (Box 2.1).

As a general rule, at present reclaimed steel sections are likely to be either cost-neutral or cheaper than their new steel equivalents for sections of 152mm (depth) and above. Smaller sections are currently more difficult to source cost-effectively. Where the cost of reclaimed steel sections is substantially above the cost of new sections, there is little point in trying to impose an environmentally 'better' solution. A number of issues need to be considered before proposing the reuse of light sections such as hand rails and other types of balustrading. This type of steel is unlikely to have identification marks and the extra effort needed in reclaiming and fabrication in a new situation may make the reclamation uneconomic.

Box 2.1 Assessment procedure for reclaimed steel sections

1. Prepare a list of the section sizes required, ideally giving a few alternatives. Also, allow for more wasted material in the initial assessment of the quantities.
2. The steel should be sourced from either demolition sites or salvage yards. It is essential that the fabricator and structural engineer visit the source of the steel to check its condition as early as possible.
3. Initial checks should take into account the degree of corrosion, which should be minimal. Delaminating steel should not be accepted. A visual survey is satisfactory at this stage, plus checks for soundness with a hammer and a screwdriver.
4. The cross-section of the steelwork needs to be measured. Components should also be checked for straightness, bow and twist. The existing finish should be noted, particularly if it is galvanized or has a special paint finish. If possible the year of manufacture should also be established from marks rolled onto the surface of the steel during manufacture.
5. Using the information obtained from the visual inspection, check the size of the steel in the historic tables to verify the dates of manufacture. The historic table then gives the stresses that should be allowed for in any calculations.
6. When inspecting the steel, any rivets, bolt-holes, plates or angles fixed to the beams should be noted as they have to be taken into account in the fabrication and the calculations. Inform the client that there are likely to be variations in the steel that you would not expect with new steel.
7. If it is not clear when the steel was manufactured and it does not have any identification marks, it may be necessary to undertake a tensile test on a sample of the steel to verify the stress that should be used in the structural calculations.
8. For any particular source of steelwork it is essential that the fabricator agrees that the steel can be fabricated for future use. This should be agreed at a joint meeting with the structural engineer. In particular the fabricator should be satisfied that cutting and welding of the steel will be satisfactory.
9. Assuming that the steel is satisfactory for use in a particular project, it should then be transported to the fabricator's works to be cut to length, grit-blasted, then painted or galvanized before erection.
10. A further check should be undertaken of the steelwork when it is fabricated before delivery to ensure that it complies with the specification.
11. The site works would be as for new steel.

Figure 2.19 Structure of new building constructed with reclaimed steel

Source: Ellis and Moore, BioRegional Reclaimed

The above procedure enables second-hand steel sections to be reused successfully in either new-build or refurbishment projects. It has to be made clear to the client that the appearance may not be the same as new steel – for example there may be holes and existing plates attached to the steel that cannot be altered without undertaking further work. If the steel is covered for fire protection this is less likely to be an issue.

Contractual issues should be resolved as early as possible so that liability and programming of the design and procurement can be organized satisfactorily. For unusual section sizes, additional time may be required to source the material over and above a normal contract where new steel is specified. Sourcing in time for construction is critical as well as establishing the length of time the steel sections have been stored. The simplest method of procurement at present is for the client to purchase the steel sections, following the above guidelines, and provide it to the contractor.

As the procurement of reclaimed steel becomes more common, there are likely to be more storage yards set up to aid procurement so that most normal sizes of reclaimed steel section will be readily available off the shelf.

Lessons learned

The structural performance of steel is relatively easy to assess, making its reuse straightforward. The economics of the supply chain, including storage, trimming, cleaning and painting will be considerably improved if the operation is organized at a regional level, rather than locally.

Structural engineer: Ellis and Moore, London
Reclamation consultant: Bioregional Reclaimed
Websites: www.bioregional-reclaimed.com
www.ellisandmoore.com

Fitout of Duchi shoe shop, Scheveningen, The Netherlands

The Dutch architectural practice 2012 Architecten has put recycling at the heart of its work, describing the methods they have developed as 'recyclicity'. This process involves an experimental phase during which design teams experiment with the materials to establish their properties and how to work and use them to make building elements. Their principal aim is to design and build buildings using recycled or reclaimed materials, not only from buildings being demolished, but from other sources too. During 2003, Architecten investigated the materials available from the car industry and found an opportunity to make imaginative use of windscreens from scrapped Audi cars when they won a commission to fit out a fashion shoe shop in Scheveningen. The centrepiece of the shop is a highly sculptural pair of sofas for customers to use. This was made from reclaimed wood and built up to create the complex form, before final shaping and finishing.

Lessons learned

When chosen carefully, reclaimed materials can be used to produce goods with a quality of finish indistinguishable from new materials.

Location: Scheveningen, The Netherlands
Date: 2004
Client: Schoenenwinkel Duchi
Structural engineer: Ove Arup and Partners
Architect: 2012 Architecten
Websites: www.2012architecten.nl/projecten/duchi.html
http://216.203.42.172:8080/recyclicity/index_en.jsp

Figure 2.20 Shelves for shoes are made from reclaimed car windscreens

Source: 2012 Architecten

Figure 2.21 Seating for customers is made from reclaimed timber

Source: 2012 Architecten

3 Making Reclamation, Reuse and Recycling Happen

Not the 'usual' approach to design and procurement

The approach needed for incorporating reused, reclaimed and recycled goods and materials is, in some ways, entirely different to the 'normal' practice of using new materials and, in other ways, quite similar. It is important to recognize these differences and similarities at the very beginning of a project. The central issue is the availability of the goods or materials (Figures 3.1 and 3.2).

Reconditioned goods and recycled-content building products

Buying goods that have been returned to a factory, reconditioned and given a new warranty of a certain performance is little different from buying new goods. Buyers simply need to make their judgement of value when comparing performance, warranty and cost with that of a new alternative. Such products are likely to be available through reputable suppliers in the necessary quantities and covered by reliable delivery terms and conditions. Among many such products are:

- many electrical goods including electric motors, transformers and switchgear;
- heaters of water and air;
- pumps and fans;
- refrigeration plant, including chillers;
- lifts.

Likewise for RCBPs. These are effectively new goods and are available from manufacturers and suppliers who make their goods available in the same ways as providers of goods made with virgin materials. Today there is a growing number of product directories dealing specifically in RCBPs (see References).

In the case of both reconditioned goods and RCBPs, the building can be designed and procured in the normal way up to the scheme design (RIBA Stage D in the UK). The type of goods to be included in the building will be established and specified either during detailed design (RIBA Stage E) or when a contractor submits a tender for undertaking the construction of the building. The building components and equipment will generally be defined by a performance specification that the contractor will need to deliver. In the case of reconditioned goods and RCBPs, the specification will be similar to goods made with virgin materials, but with the added requirement that they be reconditioned or should include a specified recycled content.

When selecting a contractor and sub-contractors to tender for building work involving reconditioned goods and RCBPs it will be important to ensure that they understand the importance of this requirement and are familiar with the market place for such goods and products.

Reusing buildings and their parts in situ

If all or part of an existing building is to be reused in situ, the following will be needed from the beginning of the project:

- availability of existing buildings for reuse in their entirety;
- full survey of the building to be reused;
- expertise in appraising the existing structure and, if necessary, defining work to be undertaken to make the structure reusable;
- expertise in appraising the existing building services, envelope and other features and, if necessary, defining work to be undertaken to make them reusable;

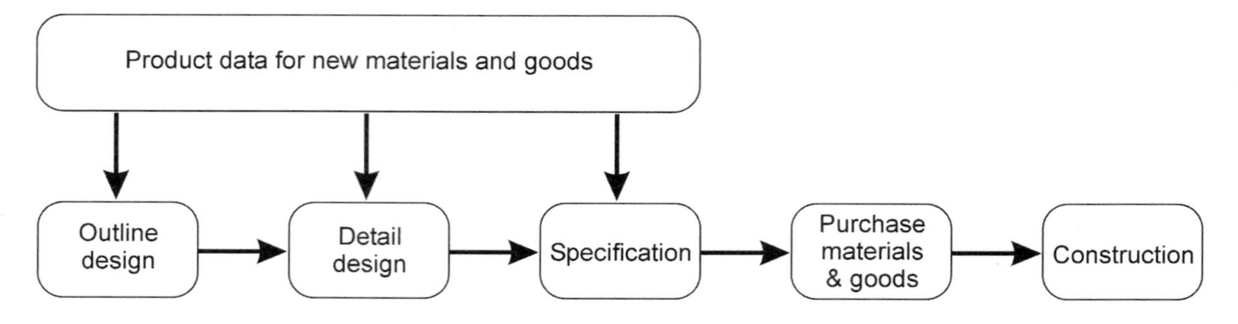

Figure 3.1 Flow chart of normal design

- expertise in adapting the existing building for the incorporation of new features (e.g. staircases, lifts, building services plant and distribution);
- if only the façade of an existing building is to be retained, expertise in façade retention;
- if only the foundations and piles of a former building are to be reused, expertise in assessing condition and capacity of piles.

Reusing salvaged goods or reclaimed materials

Incorporating salvaged or reclaimed materials and goods in new building construction requires an approach to design and construction entirely different from 'normal' building. In normal building the design team first designs the structure up to scheme or detailed design and then suitable goods and materials are sourced and purchased. In a reuse

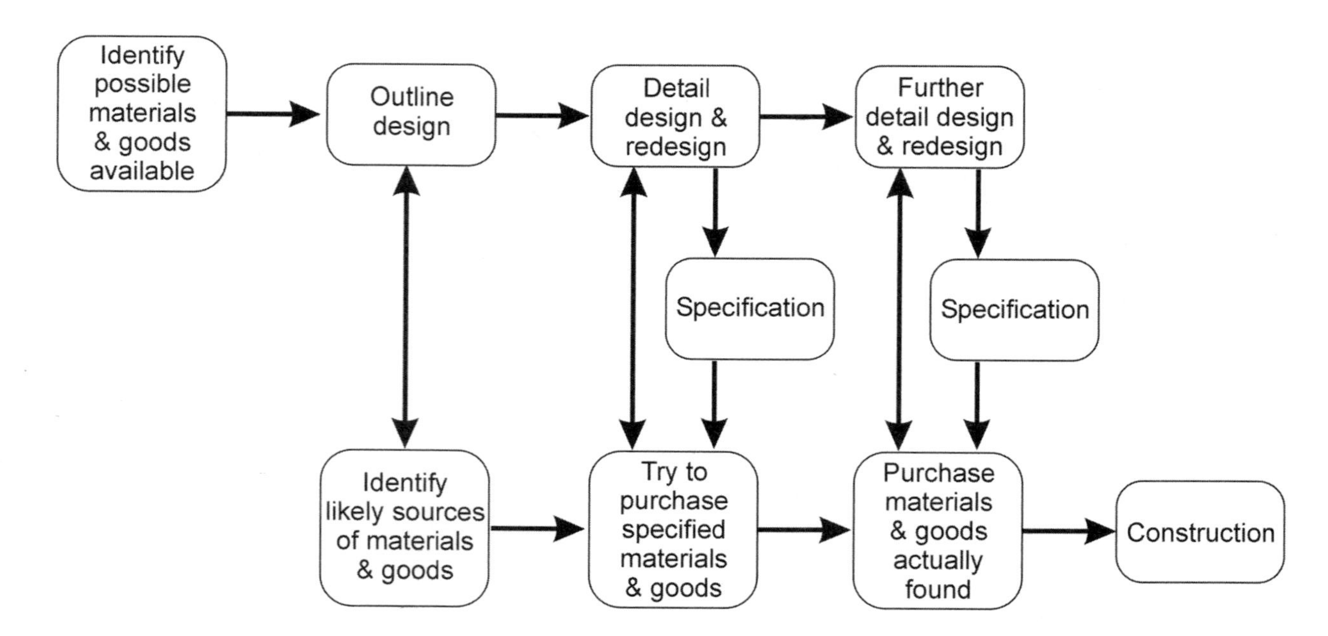

Figure 3.2 Flow chart of design with reclaimed products and materials

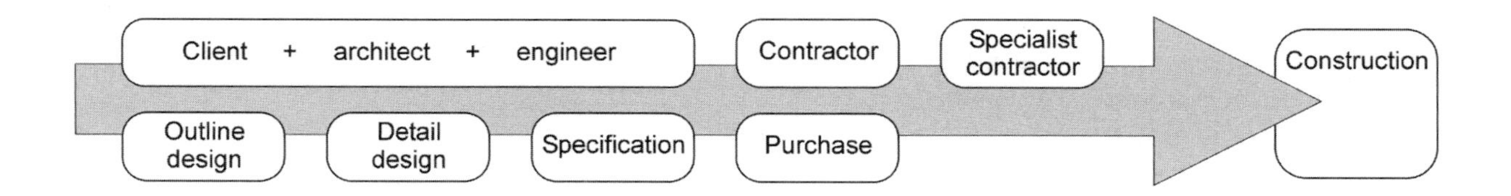

Figure 3.3 Project team for normal design

building, it will often be necessary to source and purchase the goods and materials before the design has reached the detailed design stage (Kristinsson et al, 2001) (Figures 3.3 and 3.4).

The process becomes especially complex when the design of one element affects the design of others, for example the spacing of columns may be affected by the location of existing foundations, the lengths of beams to span between columns, and the size and type of the cladding found.

Only when the items to be used have been found will it be possible to draw up a method statement and programme for preparing them for incorporation into the building. This will involve some or all of the following steps:

1. prepare an initial list of goods to be sought and identify likely sources;
2. prepare outline design of building;
3. specify approximate type/sizes/quantities/ performance of required goods/materials;
4. identify sources of the goods (in a building to be demolished, in a salvage yard);
5. assess the condition of goods;
6. develop a method statement and costs for the processes needed to bring the goods into reuse;
7. agree price and purchase goods;
8. prepare scheme design of building;
9. arrange with demolition contractor (as necessary) method for careful removal of goods from building being demolished;
10. arrange packing, transport and storage for goods, as needed;
11. arrange reconditioning/refurbishment/ remanufacture/testing of goods;
12. arrange for delivery of goods to construction site, including on-site storage if necessary;
13. complete detailed design when the precise nature and state of goods to be used are known.

In order for this process to happen, various elements are needed that are not usually required in a building project:

- availability of buildings soon to be demolished, as potential sources of goods and materials to be reclaimed and reused;

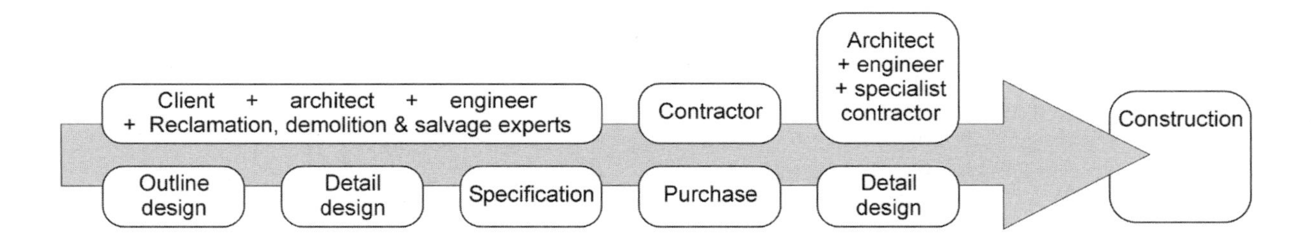

Figure 3.4 Project team for reuse design

- availability of reclaimed goods in the architectural salvage market place;
- assessments of condition of materials and products for their potential reclamation and reuse in situ or in a salvage yard, including knowledge of the processes involved (suitable contractors, duration and cost);
- suitable methods for dismantling/removal/ salvage of goods during building demolition.

Decision-making

The development of the design of any building involves a large number of decisions and, for each decision, there need to be criteria for deciding among alternatives. In the design development of a 'normal' building the criteria usually include:

- contribution to the function of the building (e.g. column spacing);
- aesthetic considerations (e.g. materials used in façade);
- technical performance (e.g. natural ventilation versus air-conditioning, acoustics);
- durability;
- construction methods (e.g. steel frame versus concrete frame);
- cost (usually capital cost only).

When designing a building with reused and recycled goods and materials, a number of additional types of decision need to be made, each with their own decision criteria.

While some of the criteria below have clear numerical measures (notably cost), others can only be measured in very vague terms and may verge on subjective choices. While one solution might be to devise and operate a scoring system to lend the appearance of rationality and objectivity, this is not recommended. In practice it would be cumbersome and a drain on resources that are probably better spent on the building itself. Among the various criteria below, many include numerical calculations that bring a degree of objectivity to the decision-making.

A sequence of decisions will need to be taken during the development of the design at three levels:

1. the building as a whole;
2. the elements (foundations, structure, etc.);
3. the individual components that are used to construct the elements.

In each case the decision will need to be taken as to the best option (best from a number of points of view) among:

- reuse in situ;
- salvage/reclaim from a building on the site, yet to be demolished (e.g. timber, crushed concrete for fill or secondary aggregate);
- salvage from a different existing building, yet to be demolished (e.g. cladding panels);
- obtain from a salvage yard (e.g. reclaimed bricks);
- obtain from a specialist firm refurbishing goods (e.g. metal window frames);
- obtain from a supplier of reconditioned equipment (e.g. electric motor, chiller);
- obtain from a supplier of RCBPs (e.g. rubber floor tiles, geotextiles, garden furniture).

Finally, it should be accepted that, for some components or materials, it will be 'best' from several points of view, including environmental, to use virgin materials or projects made using virgin materials.

Decision criteria

From a practical point of view the most important criterion is the viability of one of the reuse or recycling options – assuming it is possible, how much effort will be required, how realistic an option is it, and how easy is it to find the required items? A large number of factors may need to be considered, including:

Availability

- Will suitable products be easily available in existing buildings (prior to demolition) or in the reused/recycling market place?
- Will they be available in suitable numbers? (Compare roof tiles, identical doors, a lift.)
- Will they be available when they are needed?

- Will they be available near enough to the construction site?
- Will the goods be easy to transport and store until needed?

Ease of refurbishment

- Can the condition of the goods be easily assessed – both their technical performance and visual appearance? (Compare luminaries, a chiller, a timber roof truss.)
- Will the condition of the goods be good enough to make refurbishment or reconditioning economical? For example the cost of a new oak beam is high compared to the cost of reconditioning an old one, whereas, the cost of a new light switch is low compared to a reconditioned one.
- Will it be easy to test the goods to prove their performance after refurbishment? (Compare electric motors, windows, double-glazed units.)

Warranty, guarantees, certification, product liability and insurance

Both the project team and the building owner will require that the equipment, components and materials used in the building have some sort of guarantee as to their performance, durability and quality. If the items have been remanufactured, for example a reconditioned electric motor or carpet made from recycled polymers, they will usually be covered by a warranty or guarantee from the manufacturer or supplier.

Many recycled materials and products made with recycled materials will be required to conform to the appropriate performance or quality standards, just like new materials and products. Countries or groups of countries have their own materials standards, such as BS (British Standards) and DIN (Germany), and standards for building products (e.g. the British Board of Agrément (BBA)).

Insurance companies that provide cover for building developments generally assume and require that new building materials and products are used. Many, however, recognize that reused goods and reclaimed materials can be used. In such cases it will be necessary to discuss the proposals with the insurance company as early as possible. It is likely that insurance will be given as long as prior agreement has been obtained and appropriate and independent assessment and/or certification is provided. Major insurance companies provide technical or component life manuals that give guidance on conditions that must be met to get cover for materials and products. It is likely that an on-site inspection will form part of the assessment.

Environmental benefit

The main reason for reusing or recycling is to achieve some environmental benefit and any plans to use reclaimed goods or recycled materials must be accompanied by some means of assessing the environmental benefit of the different options under consideration. Assessing the environmental impact of construction, and hence the benefits of reuse and recycling, can be a very complicated process. A number of the different ways are discussed in Appendix B. In most building projects constraints on project costs are likely to favour using one of the simpler and quicker methods of assessment.

Diverting materials from landfill

The most direct environmental benefit comes from reusing or recycling goods and materials that otherwise would be sent to landfill sites. When selecting the reused components of a building or recycled materials to utilize it would be appropriate, therefore, to favour certain products and reclaimed materials rather than others:

- reclaiming high-density materials that would be deposited in landfill sites in large masses, for example, reclaimed bricks and concrete and masonry from demolition waste;
- reclaiming low-density materials that would be deposited in landfill sites in large volumes, for example, timber, plastics, especially expanded polystyrene and plastic bottles.

There also arises the issue of how easy it is to separate and reclaim the materials so they can be recycled. These issues are discussed in Appendix A on demolition practices.

Meeting criteria of an environmental rating methodology

Environmental assessment tools such as BREEAM, LEED and GBTool (see Appendix B) already contain precise criteria for gaining credits for reuse and recycling. It may be satisfactory to define the recycling objectives for a project in terms of achieving certain identified credits and recognized assessment tools, even if a building is not being formally assessed. For example, some credits in LEED and the GBTool are awarded according to the percentage of recycled content in the whole building. In order to establish this it would be necessary to calculate the volumes of materials and goods being reused, including the recycled content of RCBPs that is usually quoted by the supplier of such goods. It will be important to ensure the credibility of the claims made for the recycled content of RCBPs (for example LEED Credit 4.2 in Appendix B).

Assessing impact on the environment

There are many methods for assessing the impact that materials and manufacturing processes have on the environment. Some consistency of approach to such assessments has been achieved in procedures for life cycle analysis (LCA), which evaluates all the environmental burdens associated with a product, process or activity. Such analysis is complex and needs to be undertaken by a specialist in the field. To make the results of LCA more accessible to designers using new products, some guides are available that reduce the data to lists of recommended materials or products grouped in bands (Woolley et al, 1997; Woolley and Kimmins, 2000).

However, there are many difficulties in using LCA calculations for building materials and products. The results are entirely dependent on the actual materials being used – the processes used to make them and where they were made, since this affects the transport used to deliver the materials to their final destination.

A less comprehensive assessment of the environmental impact of a material or product is its *embodied energy*. This is the total amount of energy used in extracting the materials, manufacturing them into products and installing them in their final location. It is relatively easy to calculate and has the advantage that the result can be quoted in familiar units – GigaJoules/tonne (GJ/t). While this allows a quick comparison between alternative materials, the embodied energy of a given sample of building material will be very difficult to quote with confidence unless its precise origins are known. Using published data on embodied energy to attempt to make precise calculations has to be done with great care, and the methodology behind the calculation of quoted data needs to be known and made known when they are used (see, for example, Woolley et al, 1997; IStructE, 1999; Thormark, 2000; Woolley and Kimmins, 2000).

When selecting which reused components of a building or recycled materials to use based on embodied energy content, it would be appropriate, therefore, to concentrate on reusing and reclaiming goods and materials with high embodied energy. The differences between common building materials are large, for example:

- fired products such as bricks, tiles and those containing cement (approximately 3GJ/t);
- glass (approximately 30GJ/t);
- metals (45–100GJ/t);
- recyclable plastics (approximately 150GJ/t).

The use of calculations of both LCA and embodied energy for reused components and reclaimed materials brings a further difficulty – how should the data relating to the product in situ be taken into account? Apart from the fact that the precise origins of the materials are unlikely to be known, it could be argued that the energy to manufacture them has already been used. Should the assessment be taken back to the time when the virgin materials were first extracted, or should calculations be undertaken only for the period beginning just prior to demolition of the building in which the materials are found?

In practice, it is unlikely that it will be economic to undertake detailed calculations of either the LCA or embodied energy of products and materials reclaimed from buildings for reuse or recycling. Much more likely is that decisions will be based on their costs and value to the building client.

There are now a number of websites providing product information about RCBPs that include some information related to the environmental impact of the products. Typically, product search categories, the environmental rating methodology and case studies are all useful in guiding the specifier. In a product database a specifier will usually find information on the following:

- recycled materials used to make the product;
- the manufacturing process and any use of toxic materials;
- recycling of the product in both pre- and post-consumer phases;
- emissions during manufacture and in use;
- percentage recycled content;
- place of manufacture.

Common sense

While it may be considered a rather unscientific concept, the most practical approach to deciding between reuse and recycling alternatives is likely to be using common sense, based on some understanding of the nature of environmental impact and waste, moderated by cost and availability. Generally the decision for the building project team is similar to that of someone deciding whether to take their bottles to the local recycling base using a car that will cause both traffic congestion and air pollution. If just a few reclaimed roof tiles are needed, it may not be worth driving 500km to a salvage yard to look at them. Alternatively, if they are essential to the completion of a project, and no other source could be found, the decision would be different.

Cost and value

As with new construction, a major factor affecting the purchase of reclaimed components and materials or RCBPs will be their cost. Again, as with new construction, it may also be considered appropriate to compare not only their capital costs but their *whole life costs*, which will incorporate depreciation, repair and maintenance and the costs of their eventual recycling or disposal.

The costs of reclaimed components and materials or RCBPs will need to be considered in the context of the whole project. There are likely to be costs not encountered in conventional building construction, including:

- finding materials or goods in salvage yards;
- finding buildings about to be demolished that can be a source of useful components and goods;
- careful deconstruction (compared to normal demolition practices);
- storage immediately following deconstruction and after refurbishment for reuse;
- refurbishing goods or materials for reuse;
- testing materials and goods to prove their performance.

The prices of reclaimed materials and goods depend, of course, on their abundance or scarcity and they can be highly volatile – one week reclaimed timber may be available at little more than the cost of transport, another week it may cost a lot more.

It is worth noting that there is often some correlation between cost and embodied energy since it is often the cost of the energy of processing or manufacture that makes some materials expensive, for example plastics or aluminium. The value of a reclaimed product or material can thus give a rough indication of its environmental benefit.

In some cases, such as with demolition waste, there can be several cost benefits to recycling. Demolition waste can be crushed on site and used as backfill, for landscaping or the sub-base for an access road or car park. This will avoid the cost of removing the waste from site, the cost of depositing it in a landfill site and the cost of buying in material. Demolition waste can also be crushed and used as a secondary aggregate for low-strength concrete, in which case the cost of buying new aggregate (including any aggregate tax) will also be avoided.

Generally, it will not only be the cost that affects purchasing decisions because some consideration of environmental benefit will need to be factored in. This effectively means the decision will be based on *value* rather than cost. Unlike cost, value is often subjective. In the case of a building made with reused and recycled goods, this will reflect the precise nature of the objective or commitment to reuse/recycle and the degree of determination shown by the client and the project team to achieve the objective.

Decision sequence

The full decision-making process for acquiring and incorporating reused or reclaimed goods and materials will comprise a number of stages. These may apply to the building as a whole and/or to each building element, component and material:

1. discuss and agree, as far as possible, the reuse/recycling objectives for the project, and the strength of commitment to achieving them;
2. discuss and agree, as far as possible, the decision criteria to be used for material/product selection;
3. set targets for the level of reuse/recycling (reuse in situ, RCBPs etc.);
4. identify those components/materials with no recycled content;
5. assess possible sources of goods/materials for viability, environmental benefit and cost;
6. assess the condition of reclaimed elements/components/materials, and remedial work needed to raise them to the necessary quality for reuse;
7. evaluate costs and environmental benefits and purchase.

Opportunities for reuse and recycling

Reusing and recycling materials

Materials are used by the designers of buildings in one of three different ways: as materials in themselves, such as concrete or timber: as components and products that are made from materials, such as a window frame, boiler or electric motor: or as a hybrid between the two, part as material and part as product, such as a steel beam or timber rafter.

In the first case, the designer will need to know the engineering properties of the materials, such as strength, thermal conductivity or resistance to corrosion. In the second case, the building designer will be interested only in the performance of the component or product as a whole, such as the heat output of a boiler or the functionality of a window. In the third case, which concerns mainly the load-bearing elements of buildings, the performance of the product can only be assessed by considering both the product as a whole – the beam – and the engineering material(s) of which it is made – steel or reinforced concrete. There can be some ambiguity about what constitutes a material and what a product. Bricks that can be separated and reused as bricks are considered as products; a timber beam reused as a timber beam is a product. Brick that is crushed and used as secondary aggregate is a recycled material; so too is a piece of timber that is processed into a different form such as granules or very small pieces, and used to make a product such as blockboard or chipboard.

These distinctions are important when we look at how materials can be reused as artefacts or recycled and made into new products. Of similar importance are many of the practicalities of demolition processes and the handling of the material streams that result from demolition. These are reviewed in Appendix A.

Most materials can be reused and recycled as long as their condition and properties are known and judged to be adequate for the purpose intended or can be treated to improve them to an adequate level. Whether this happens in practice depends entirely on the commercial viability of the various processes (Gorgolewski, 2000; Kristinsson et al, 2001).

Reuse

If a beam made of timber is to be reused as a beam, it will be necessary to assess the quality of the material and whether the beam as a whole will perform satisfactorily as a beam. The opportunities to reuse materials in the form of components will depend on two main factors:

1. how easily the properties, performance, condition and quality of the materials to be reused can be assessed in order to:
   - design a component using the material (e.g. a floor beam), and/or
   - establish the life or durability of a component made with the material;

2 the nature of remedial work or reconditioning that can be done to improve the properties, performance, condition or quality, to a level that enables it to be reused. (e.g. removing nails from a timber floorboard.)

The term 'reuse' tends to imply that the material or product is being used for the same purpose as it was used in its former life – a steel beam being used again as a beam. This need not always be the case; a piece of oak formerly used as a floor beam could be used to repair a roof truss.

Recycling

When materials are recycled their new life begins as a raw material for a manufacturing process and the primary concern is knowing the ingredients or chemical composition of the material. For this reason, an essential requirement for recycling is that the material must be as pure, consistent and uncontaminated as possible, for example crushed masonry containing no timber or plastics, plastics all of the same chemical type (e.g. polypropylene), or wood chips with no plaster or plastic.

With the exception of crushed aggregate being used to make concrete, the recycling of materials does not take place on the construction site. Timber, plastics and metals are all recycled in specialist factories or industrial works outside the construction industry. In the case of recycled materials there need be no link between the first and second uses. Timber window frames could be recycled as chipboard and used to make kitchen cabinets; plastic drinks bottles can be recycled to make plastic drainpipes and car tyres used to make acoustic insulation mats. Materials reclaimed from a demolished building or from construction waste may be recycled and used outside the construction industry, and the recycled materials used in recycled-content building products may have been reclaimed from outside the construction industry.

The opportunities to recycle materials depends on two main factors:

1 Whether the material can easily be separated from other materials – iron and steel, for instance, can be separated magnetically; copper and tin cannot.
2 The suitability of the recycled material to be manufactured into a useful product, for example recycled plastic bottles used to make garden furniture. Such recycling often involves mixing a proportion of recycled material with virgin materials. It should be noted that some materials, such as aluminium, can be recycled almost indefinitely since the reprocessing (remelting and assaying) results in virgin aluminium. Other materials, such as plastic or timber cannot be recycled indefinitely – after each recycling, plastic becomes less suitable for manufacturing processes such as injection moulding; similarly, timber recycled as chipboard or MDF, becomes contaminated with the phenolic resins used to bind the particles.

Exemplar buildings

The most effective way for any new type of construction to be demonstrated is for an exemplar to be built. Since such projects can be more expensive than conventional construction, a valuable role can be played by national and local governments in funding such experiments in order that the lessons learned can be passed on for others to learn from. In the UK both the 'Cardboard School' and the 'Reclaimed Vicarage' (see Chapter 2) were government-funded exemplar projects.

The RecyHouse, constructed on the site of the Belgian Building Research Institute (BBRI), is another example, in this case funded mainly by the European Commission. This building was conceived to incorporate a large proportion of new construction materials produced with recycled materials of all sorts. The objective was to demonstrate that it is indeed possible to construct a building almost entirely with recycled materials. The main goal of the RecyHouse was to inform all those associated with the building process about the possible applications of recycled products.

Location: Limelette, Belgium
Date: 1996–2001
Client: Belgian Building Research Institute
Sponsorship: European Commision
Architect: Jacques Willam
Structural engineer: S.E.C.
Ref.: 'Opportunities for using recycled materials in the construction sector', 2002, available via the website
Website: www.recyhouse.be

Opportunities for reusing and recycling materials

While the following seven headings in this section cover the main materials used in construction, it would be fair to say that almost any material can be recycled in some way. The recycling of materials generally falls into three kinds. The first is finding uses for materials that would otherwise be called 'waste' and have to be disposed of in some way – materials such as waste from mines, dredging, incinerator ash, sewage sludge, and so on. These have very limited application in building construction though are widely used in civil engineering projects. The second type of recycling is of materials that have been produced as a by-product or waste from an industrial processes. Finally there is the recycling of post-consumer waste that has to be collected and separated from other materials to make it suitable for remanufacture.

The circumstances in which recycling may be a practical proposition vary from material to material and product to product (Rayner, 2002). A great many manufacturers recycle post-industrial waste as a normal part of their process, and have done since the dawn of manufacturing. Indeed, it could be argued whether this sort of recycling should really count as recycling, though it is unquestionably something that should be done. For example, waste materials created in the production of plasterboard can easily be recycled in the factory. It is a different sort of recycling when surplus off-cuts of plasterboard are returned from the construction site to factory to re-enter the manufacturing process. This is now being done by some contractors. A further step is for plasterboard removed from buildings during demolition to be returned to the factory and the gypsum recycled. Whether this can be done depends on how clean and uncontaminated the material is.

Soil and excavation spoil

While the material on which buildings are usually constructed is often called 'soil', it may vary in its make-up from solid rock to an almost-liquid clay, and its constituents and their proportion are likely to vary with depth. Soil needs to be assessed for use or reuse in three main ways – its load-carrying capacity, its permeability to water and its ecological properties and suitability for growing plants.

Assessing the suitability of the soil for load-bearing purposes and its permeability to water, whether in situ beneath the building or brought in as fill from elsewhere, is the work of the ground or geotechnical engineer. Such tasks come in the normal scope of their work. It is usual to assess the engineering properties of soils in the site investigation undertaken prior to any excavation or construction. Samples of soil are taken from boreholes or trial pits and sent to laboratories for analysis. The interpretation of the results is undertaken by the geotechnical engineer.

Considerable quantities of soil, including naturally occurring material, rubble and demolition waste used as fill and topsoil, are excavated during construction work and much of this is often sent to landfill sites. Landfill taxes have encouraged construction firms to look carefully for uses for material excavated from sites, but there remain many occasions where such material can be reused for landscaping or as fill.

Despite the large quantities of material involved, the reuse or recycling of soil is seldom given the attention it deserves in guidance to building designers or landscape architects.

Masonry and fired clay

Masonry and fired clay comprise both natural materials – stone – and artificial materials – bricks, blocks and tiles made by firing clay, and blocks and tiles made from concrete. Stone, brick and blockwork is usually constructed using a mortar to provide good bedding between units and, to an extent that varies with the type of mortar, to bind the units together. All these materials are highly durable and can last well over a thousand years.

In Britain alone over 2.8 billion new bricks were made in the early 1990s, all made from virgin clay that is a non-renewable resource. The energy used to fire bricks and tiles at over 1000°C is also considerable. These two facts mean that there is great environmental benefit in reusing bricks and tiles if possible.

The ease with which stones, bricks and blocks can be separated for reuse depends on the type of mortar used. Modern cement mortars are highly tenacious and make separating the units both mechanically difficult and likely to cause damage to the units. Before the early 20th century, lime mortar was used for stone, brick and blockwork. This bonds much less strongly to the units making their separation and reclamation much easier – a rate of up to 2000 bricks a day can be achieved – and, hence, commercially more viable.

Although bricks are usually made from virgin clay it is possible to make bricks with recycled content using a variety of post-industrial waste materials, including colliery spoil, dredged silt, pulverized fuel ash (PFA), blast-furnace slag and sewage sludge. Concrete blocks too can be made using a proportion of reclaimed or recycled materials including PFA and blast-furnace slag (BS6543:1985).

Further guidance on reuse of masonry construction and the use of recycled-content bricks and blocks is given in Chapter 5.

Timber

In Britain alone over 50 million cubic metres of timber were used in the UK construction industry in the early 1990s and concern was growing worldwide that sources of timber were not being replenished by new planting. This led to the establishment of the Forest Stewardship Council (FSC), which gives accreditation to those suppliers who do plant trees to replenish supplies.

Relatively little attention has been paid to the reuse of timber as another way of reducing the environmental burden of using timber. Over half a million tonnes of timber arises from demolition each year in the UK alone, of which only about 30 per cent is reclaimed and reused or recycled. About 6 per cent is burned to generate energy and about 1 per cent is recycled as bedding material for pets and for use in gardens (Kay, 2000).

Timber is used in a wide variety of construction components and building elements and the material is used in many different forms, varying from substantial structural timbers that may be many hundreds of years old, to modern products such as chipboard and medium-density fibreboard (MDF), which are made from small particles of timber bonded with a resin glue. The function of timber products also ranges widely, from substantial beams and roof trusses to finishing elements, such as picture rails. The large quantities of timber used in the construction of buildings, and the ease with which its condition can be assessed make it a good candidate material for reuse. The use of timber to make formwork for concrete accounts for a significant proportion of all cut timber and timber panel products (such as plywood) used in building construction. Using reclaimed, in preference to virgin, material would make a noticeable impact on the use of new timber.

The opportunities to reuse timber in construction vary greatly according to the type of timber product and its intended use. Softwoods are highly susceptible to damage in the deconstruction or demolition process, either through the breaking of slender lengths of timber or surface damage and indentation. Nevertheless, reclaimed timber does present many opportunities for reuse and recycling, depending on its form:

- sold by length or volume for reuse as structural or non-structural timber;
- reuse for making formwork and shuttering in concrete construction;
- recycled to make chipboard for use in furniture or kitchen manufacture;
- recycled as wood chippings and used as a soil improver.

Old timber may often be of superior quality to modern timber with fewer defects, a tighter grain, well-seasoned and available in sizes, lengths and species that today may be difficult to produce from sustainable sources (Ross, 2002: Yeomans, 2003).

The ease with which timber can be reclaimed depends on its size, how it is fixed and what is fixed to it. Large structural timbers are usually easy to remove without damage. Timber studding (typically around 75 × 50mm) needs to be removed with care to prevent damage and will require the removal of nails and cleaning to remove plaster that has adhered. Timber in smaller sizes and most timber panel products are more difficult to remove without damage and, when they are reclaimed, are usually sent to factories for reducing to chips for making timber panel products.

Whilst there is a ready market for clean, used timber, contaminants that can easily become mixed with the load will result in the timber being rejected as a recyclate. The effort required to selectively separate timber from all contaminants may be deemed too expensive to justify the returns.

The storage of timber between reclamation and reuse is an important issue. Stockpiles of timber stored on site can occupy a large area that may cause problems when space is restricted on a small site. Also stockpiles on site may constitute a significant fire hazard, and careful storage will be needed to minimize any risk.

There is a growing market for chipped timber – already over 100,000 tonnes per year in Britain. However it is highly sensitive to market forces – as supplies increase, demand can quickly be satisfied resulting in a rapidly falling price for the raw material. The waste timber is separated from other waste streams and collected from demolition and construction sites. After delivery to factories where it is reduced to chips of various sizes it is used to make a range of 'forest products' including chipboard, MDF, hardboard and so on. Some materials (e.g. MDF) can only be made from post-industrial waste, others from post-industrial or post-consumer waste. The environmental disadvantage of this process is the relatively high environmental impact of the resins used to bond the wood particles. Such forest products are used mainly for non-structural purposes.

Though beyond the scope of building construction, it is worth noting that waste timber can also be pulped to make paper and board. Also, although a lot of waste timber is still burned with no thought to the environment, it is increasingly incinerated to liberate energy that is used to heat water or to generate electricity.

Metals

All metals require a great deal of energy in their extraction for ores and their manufacture into artefacts. This results in high prices compared to non-metallic materials used in building construction. Metals are often easy to separate from each other and from other materials. Some metals are easy to separate from mixed waste – iron and steel can be removed with electromagnets and aluminium and copper can be removed using other electromagnetic processes. Finally, metals can often be separated from each other and other materials using a fluidized bed that segregates materials according to the densities. Finally, metals are easy to recycle by simply adding them to the melt in the furnaces where metals are made. For these reasons only a small (but still significant) quantity of metal finds its way to landfill sites; the metal that does is usually bonded very firmly to other materials with very low value that makes separation uneconomic, for instance some composite panel products and light fittings. In the re-smelting process, the properties of metals are fully restored, though not always easily, regardless of their physical or chemical form. The resulting metal is effectively new material and this means that metals can be recycled almost indefinitely.

Metals are usually considered under the two main categories of ferrous metals (iron and steel) and non-ferrous metals.

Ferrous metals

Ferrous metals include many hundreds of different alloys of iron including wrought iron, cast iron, mild steel, stainless steel, weathering steel (e.g. Cor-Ten), high-tensile steel, and so on. Different alloys comprise at least 80 per cent iron with up to about 5 per cent carbon. The remaining proportion consists of other elements such as copper, chromium, manganese, molybdenum, tungsten and many more.

Until the 1770s the only alloy of iron used in buildings was wrought iron that continued in use until the 1890s. From the 1770s to the 1880s cast iron was used for structural and decorative purposes. From the 1880s steel quickly replaced wrought and cast iron for all structural purposes. A half dozen or so new alloys such as stainless steel (including chromium) and weathering steel (including copper) were introduced into the building industry during the 20th century. (The remaining dozens of iron alloys are not used in the construction industry.)

In general the steel used in the frame has high reuse/recycling value. One reason is that structural steel is made in standard sections and a number of standard grades (yield strengths). This means there are likely to be many potential uses of a second hand I-section beam. Also, steel sections are very versatile and structural beams from buildings may be reused in civil engineering projects and temporary works; industrial buildings and their sheet-steel cladding can be used in agricultural buildings. In these cases, section sizes and the presence of holes in beams can be tolerated as large safety margins can be used and aesthetics are seldom important.

The robustness of steel products means that many can have a long life and be reused many times. Sheet-steel piles, for example are often removed and reused. The exception is when concrete has been cast against them or they form part of the permanent works.

There is a well-established recycling market for most steel goods. The scrap value of iron alloys varies according to the particular alloy. Ordinary mild steel scrap was fetching as little as UK£2 per tonne (in 2003), steel reinforcement between £10 and £30 per tonne and stainless steel a little higher. All prices are highly dependent on market conditions.

There are clear environmental benefits in reusing steel beams and columns since energy is saved twice – first in the energy that would be needed to re-melt the steel in a furnace, and second in the energy saved by not needing components made from new steel.

Concerning the use of recycled steel, the matter is more complex and, to some, controversial. Most steel worldwide is made using one of two industrial processes. Global steel production in 2002 was around 902 million tonnes. Of this total, 60 per cent (541 million tonnes) was produced by the *basic oxygen steel-making* (BOS) process, which uses iron ore and up to 20 per cent scrap steel. A further 34 per cent (306 million tonnes) of the world's steel was produced by the *electric arc furnace* (EAF) process, which involves melting up to 100 per cent steel scrap. The remaining 6 per cent of production is made using the open hearth and other production methods (Steel Construction Institute, www.steel-sci.org/).

It can be argued that, in the long term, specifying steel with recycled content will tend to increase the proportion of EAF steelworks. However, the world-wide demand for steel and production of new steel considerably exceeds the amount of scrap iron and steel that is available and virtually all iron and steel is already recycled at the end of its first life and returned to steelworks to make new steel. New steel from made from iron ore is, therefore, still needed in large quantities.

In the short term, and for a particular building project, it would thus make no difference to world steel production for a building designer to specify 'recycled' steel. In practice, this would simply constrain the contractor to purchasing steel from certain sources, while the previous buyers of steel from those steelworks would be forced to buy steel from other works that process scrap iron and steel.

When considering recycled steel, it is important to note that on a global scale the recycled content is irrelevant providing that the steel is reused or recycled in a similar grade application at the end of its life. Global demand for new steel exceeds the supply of scrap steel by a factor of around two and therefore, to meet this increasing demand, new steel has to be produced from primary sources.

If designers specify steel with a high recycled content, the result in some geographical regions could be an increased demand for scrap steel that would drive scrap prices up to unreasonable levels and would raise the cost of steel made by the EAF process. It would also probably result in increased environmental impacts from transportation, as 100 per cent recycled steel may not be manufactured locally.

Non-ferrous metals

Copper, brass, bronze, aluminium, zinc, tin and lead tend to have a higher value than ferrous metals (up to UK£1500 per tonne) and so have a great potential for recycling. However, a similar story applies concerning the use of 'recycled' non-ferrous metals as applies for steel – relatively little metal is discarded as waste and the demand for new metals exceeds the supply of used metal. There is still, therefore, a demand for virgin metal, even if it is mixed with recycled metal.

A key difference between ferrous and non-ferrous metals is that the latter are not made in standard sections with the exception, perhaps, of copper pipes and electric cable. Indeed one of the main features of most metals is the ease with which they can be made into shapes and sizes that ideally suit the purpose at hand, for instance, window frames, pipe fittings, pressed façade panels, connectors in electrical equipment, door and window fittings, and so on. While entire products containing such unique metal components may be reused (e.g. electric motors), it is highly unlikely that the metal components themselves would be reused.

Concrete

Concrete consists of aggregate (typically around 55 per cent gravel and 25 per cent sand), cement (14 per cent), water (6 per cent) and various additives to influence its viscosity when pouring and the chemical reaction during its curing. Its use has considerable environmental impacts – especially the quarrying or off-shore dredging for gravel and sand, and the large amounts of energy used in heating the lime to around 2000°C to make the cement. After use in a building, much concrete is (or used to be) crushed and sent to landfill sites.

Although concrete may have a limited economic value as a reusable waste product, its widespread use in construction means that it is a good candidate for reuse or recycling in order to minimize the costs and environmental impacts associated with its disposal (Bitsch, 1993; de Vries, 1993). Taxes on landfill and the extraction of virgin aggregate have been introduced in many countries to influence the market and encourage a reduction in the use of virgin aggregates and the reuse of crushed concrete arising from demolition.

Five options are available for reducing this environmental impact:

1 prolonging the service life of concrete structures;
2 reusing concrete components reclaimed from buildings after deconstruction;
3 replacing virgin aggregate by other materials;
4 replacing lime-cement by other materials;
5 finding uses for crushed concrete from demolition waste.

In situ concrete elements such as beams, columns, foundations and slabs are all designed precisely to suit their structural duty. Only in the form of standard components, such as precast planks, beams and frames, is it likely that concrete components will find uses in new situations. (Steel, by contrast, usually comes in the form of standard sections.)

Precast concrete floor panels, usually forming part of a proprietary system, can be separated from the structural frame and can be lifted from the frame. Although usually crushed at ground level they have potential for reuse provided they are not damaged and a similar structural grid is used.

Unlike steel, which can easily be recycled by heating, the creation of concrete is a one-way, irreversible chemical process. The only potential for recycling is to crush concrete and use it either directly, for fill or landscaping, or indirectly as secondary aggregate for new concrete. The raw material composition is such that recycling of pure concrete is not problematic. However, this is not necessarily the case for concrete that has become contaminated during use or has been subject to chemical attack, for example concrete used in chimney stacks in industrial manufacturing. Large volumes of concrete are crushed for reuse and the steel reinforcing bars separated for recycling. Its bulk, low value and the abundance of its raw constituents mean that it is generally not economical to transport it over great distances.

The major use of recycled crushed concrete as an aggregate replacement (recycled aggregate or RCA) in buildings is for making low-strength in situ concrete, typically replacing 20 per cent of the gravel aggregate, such as for concrete slabs for the foundations of houses and ground-level car parking areas. RCA can also be used to make precast concrete blocks and other lightly loaded units. Pulverized fuel ash can be used to replace around 20 per cent of cement used in concrete (Collins and Sherwood, 1995; Collins et al, 1998; BRE, 1998).

Further details of recycled-content concrete are given in Chapter 5.

Glass

There are three general types of glazing that can be a source of glass for reclaiming and reuse – panes from single-glazed windows, sealed-unit double-glazing, and glass panels made of toughened or laminated glass from façade or cladding systems. Apart from the danger of damage, the most difficult hurdle to overcome is the requirement for increased thermal and acoustic insulation in the building envelope. This means that single sheets of 4–6mm glass are now unlikely to meet thermal performance requirements. Sealed double-glazed units can be dismantled, cleaned and reassembled achieving improved insulation, if necessary, with a larger void between the two panes.

A significant barrier to reusing sheet glass is the potential for accidents during its removal. Often the quantity of glass removed from buildings for reuse or recycling proves insufficient to justify the cost of a segregated material stream, and it currently rarely proves to be cost-effective to recycle waste glass. This may change when buildings that have followed the recent fashion for huge glass façades come to have their façades replaced or when they are demolished.

Glass is one of the easiest materials to recycle, though remelting it is an energy-intensive process. Also, there are many different uses for recycled glass. Today the majority of glass that is recycled is crushed and used in the manufacture of new glass containers or fibreglass insulation. Only a few such products are used in the building industry.

Finely-graded glass has also been tested in the late 1990s in the US in elastomeric roof coatings, a blend of polymers and fillers spread on roofs to minimize the effects of weathering. Another potential benefit of using glass in roof coatings is as a replacement for industrial mineral fillers that are currently used, such as flint, talc and various dry clays. These contain naturally occurring crystalline silica whose dust has been associated with serious lung disorders, such as silicosis. Whilst glass is produced from silica sand, the manufacturing process converts the crystalline structure to an amorphous state. Tests have shown that recycled glass contains less than 1 per cent crystalline silica, much less than the materials it replaces.

The chemicals used in the coating of some high-performance glasses that are used in modern façades prevent them from being recycled in the melt to make new glass. Laminated glass may also be difficult to recycle since it contains layers of polymeric adhesive that bonds the separate layers. Advice on the recycling of glass must be taken from glass manufacturers.

An innovative application for waste glass in new construction is its incorporation within asphalt, to make a product known as 'glassphalt'. This was developed in the late 1960s in the US as a means of disposing of waste glass. Glassphalt is similar to conventional hot-mix asphalt, except that 5 to 40 per cent of the rock and/or sand aggregate is replaced by crushed glass (cullet). The cost-effectiveness of substituting glass for conventional aggregate is highly dependent on the location of the waste production and the quality and cost of local aggregates.

A great deal of experimentation is under way to find other uses for crushed or pulverized glass. These include its use as a micro-aggregate in concrete, and as a filler-material for bricks, tiles and clay drainage pipes. A particularly interesting application being developed is the use of glass in conjunction with a foaming agent to make blocks similar in size to concrete blocks, but lighter and with better insulation properties.

A few architects have approached the challenge of using recycled-content glass from another direction. They have found a ready source of sheet glass in the windscreens of vehicles (found in scrap yards) and designed a façade system around the product. While this may not find mainstream use, it illustrates the fruits of an original and imaginative approach to recycling (see Chapter 6).

Plastics

The construction industry is a major user of plastics, accounting for around a quarter of annual consumption, second only to the packaging industry. Although some plastics cannot be recycled (because of their chemical composition), many can and the plastics industry is already well along the road to having a developed recycling sector. Trade associations such as the British Plastics Federation publicize information about recycling and operate a materials exchange to help match supply and demand.

There are two powerful reasons for recycling plastics. They require a high amount of energy per kilogramme in their manufacture – three times more than steel and 50 per cent more than aluminium. Plastics also constitute a 'waste problem' – they occupy a large volume and the majority of plastics are not biodegradable.

A certain degree of confusion about recycling polymers arises from the fact that a recycling symbol (the three arrows) appears on the bottom of most polymer materials, thus making it seem as if they are all recyclable. In theory this may be true; but in reality very few types of polymer are commonly recycled.

Successful recycling of polymers depends on efficient separation of polymers according to their type. Different resins usually need to be reprocessed or recycled separately and, even within one type or category, different colours or pigments can make recycling more difficult to achieve. To aid the separation of post-consumer waste for recycling purposes, seven basic kinds of polymers are distinguished by the following code numbers that appear inside the recycling symbol:

1 polyethylene terephthalate (PET), for example plastic bottles, geotextiles;
2 high-density polyethylene (HDPE), for example underground drainage pipes, 'plastic lumber';
3 polyvinyl chloride (PVC), for example plastic water pipes;
4 low-density polyethylene (LDPE), for example packaging, insulation, pipes;
5 polypropylene (PP), for example plastic water tanks, drainpipes;
6 polystyrene (PS), for example plastic cabinets for computers and electronic equipment, and insulation and packaging as expanded polystyrene;
7 others.

Of these seven types of polymer PET and polyethylene are the easiest materials to recycle. PVC is widely used in construction. However, environmental concerns relating to its life cycle have led to increasing use of alternatives such as PVC-free linoleum, HDPE and vitrified clay pipes for drainage, and aluminium and timber for window frames. Polypropylene currently has limited recycling potential. Type 7 or 'others' is any other type of polymer and may be formed from different types of polymer, which makes recycling impossible. Recently some US manufacturers have developed recycling technologies for polymers that include the use of mixed polymer wastes including polyethylenes, wire strippings, carpet and flexible vinyl in making injection moulded flooring products.

A great many products are available for use in the building industry that are made entirely or largely of recycled plastics (Kilbert, 1993; BRE, 1997). Furthermore, they are relatively easy to find. The plastics industry has encouraged manufacturers to publicize the recycled content of goods containing recycled plastics and many databases featuring such products are now available via the internet. Although the title RCBP applies to any product with recycled content, the majority of such products are ones that contain recycled plastics.

Among the enormous range of products made from recycled plastics currently available are:

- geotextiles and earth retention products;
- insulation materials;
- window frames;
- roofing materials;
- various types of pipes and ducts;
- panel products for making kitchen worktops and cupboards;
- carpets, tiles and other floor coverings;
- street, park and other outdoor furniture and fittings made from 'plastic lumber';
- synthetic surfaces for playgrounds and sports fields;
- acoustics screens and barriers made from recycled car tyres.

Opportunities for reuse and recycling, element by element

Building components and equipment are only likely to be reusable if they can easily be removed from a building without being damaged. This will depend on both the methods of fixing and the sequence in which various components are assembled into a building and would need to be disassembled.

Some equipment is already made and installed in ways that are easily reversible. This tends to depend on when during building construction something is installed and whether it is felt likely that it may need to be removed for repair or maintenance – lift motors, for example. Other components, such as many cladding systems, are sometimes installed so they can be easily dismantled, and sometimes not. To some extent this depends on whether the system has been designed with deconstruction in mind.

Someone with a good knowledge of building construction and demolition practices will be needed to assess the ease with which particular equipment or components can be removed, most likely a demolition contractor. Someone with different skills will be needed to assess whether the equipment or components can be economically refurbished and reused. Whether this happens will depend on whether someone has indicated they want it to

Table 3.1 Sample table

	Reuse in situ		Use of salvaged/ reconditioned products/ reclaimed materials		RCBPs	
Element/component/product	Viability	Environmental benefit	Viability	Environmental benefit	Viability	Environmental benefit
Piles						
Timber piles	M	H	L	L	–	–
Concrete piles	H	H	L	L	L	L

happen, and whether it is possible to refurbish the goods to bring them back into use.

When looking for opportunities to specify reuse elements, components or products in a building, or to specify products made using recycled or reclaimed materials, it is important to quickly find those opportunities that are likely to be relatively easy, cheap or viable to achieve. The following tables will help the reader identify the 'quick wins', and where to find further details and examples elsewhere in this book (Figures 3.1–3.6; Box 3.1).

It should be noted that for every type of reuse or recycling it will be necessary to ensure the following:

- an adequate supply (quantity, quality and delivery times);
- an appropriate warranty of performance and/or durability, either from a manufacturer (for example a reconditioned boiler) or from a member of the design team (for example the strength of a reused oak beam);
- a contractor who will be willing and able to deal with products or materials that are not standard, familiar or new.

Seeking and finding the buildings, goods and materials

A large proportion of building materials and components can be obtained through the reconditioned, reclaimed and recycled market. Whether these are or can be used in a particular building project will depend very much on both the determination and the patience of the client and project team. Finding suitable goods and designing the building to enable them to be incorporated will require more effort and a greater degree of adaptability than is normal when designing and constructing a building. For each material or product the following key issues need to be addressed when seeking the goods and comparing the various alternatives that are found:

- performance, quality and durability;
- installation procedures;
- guarantee or warranty provided by supplier/manufacturer;
- certification by an appropriate organization;
- insurance and product liability;
- cost and value-for-money;
- precise wording to be included in specifications;
- procedure for purchasing, including guarantees on price and delivery.

In addition to these there will be two further issues that would normally not need to be addressed:

- assessing and comparing the environmental benefits of using the reclaimed materials and goods;
- a 'reality check' comparing the recycled options with normal, virgin-material options – a reuse or recycling option may lead to greater environmental impact than the virgin-material option.

Box 3.1 Key to the tables

Key to column headings

Reuse in situ

The element, component or material will be reused in its original place in a building, or with minor relocation, but not removed from site. Its performance and/or suitability can be assessed in situ to allow designers to predict performance and durability with confidence. For example reuse of structural frame or roof truss of an old building, reuse of piles after a building has been removed, or reuse of a lift after refurbishment in situ.

There is probably some ambiguity about what constitutes 'continued use' and 'reuse' of a building. General repair and maintenance apply to the former but not to the latter. For major items such as the building structure, foundations and envelope two factors need to be considered: Is there a change of ownership? Is there a significant change of use? If the answer to either is 'yes', then it should be considered to be 'reuse'.

Use of salvaged/reconditioned products/reclaimed materials

An element, component or material can be specified that has formerly been used in a different location. According to the item, it may simply have been removed from another building, stored and delivered to a new site (for example an ornamental item salvaged from a building) or it may have undergone some repair, reconditioning or minor remanufacture before being purchased for reuse, for example a window, a heating boiler, water pump or a timber beam cut to suit a new purpose and location. This category of goods includes many items that are generally found in architectural salvage yards.

RCBPs

This includes all goods that have been made using some reclaimed materials. These may be post-industrial waste – the waste material from manufacturing processes such as carpet manufacture or carpentry. Or they may be post-consumer waste – material that has been used and discarded as waste, such as plastic drinks cartons or timber stripped out when a building is demolished, for example furniture made from recycled polymers (plastics), timber products such as chipboard, concrete made using a cement replacement such as PFA, or aggregate made from crushed concrete or masonry demolition waste.

Note: This excludes metals (steel, copper, aluminium etc.) since the percentage of recycled material in a batch of 'new' metal varies with the producer of the metal. Specifying steel with a high content of recycled material (feedstock) will affect only where it is purchased, not how much recycling of steel goes on in the world.

Viability

Three possible levels of viability are indicated in the tables. The rating attempts to give a common-sense amalgamation of many disparate issues. It aims to identify those elements of buildings where 'quick wins' are likely, and those where they are less likely – it does not mean that reuse of an element is already widespread.

H – there is a high likelihood of reuse in situ or from another location:

- reuse is already well established;
- technical procedures or standards already exist;
- there are few technical barriers to reuse;
- supplies are plentiful;
- costs are reasonable.

M – there is a medium likelihood.

L – there is a low likelihood that the element or product will be reusable.

Note: The life of the element, component or equipment can always be extended by normal maintenance, repair or refurbishment, for example redecoration, repair of carpets, refurbishment of window frames, maintenance of a heating boiler, and so on.

Environmental benefit

H – there is a high environmental benefit of reuse that:

- diverts large quantities of material (by mass or volume) from landfill;
- retains in use equipment, components or materials with a high embodied energy.

M – there is a medium environmental benefit.

L – there is a low environmental benefit because only small quantities are used.

In the tables below 'quick wins' can be identified as HH – those options that are both viable and bring significant environmental benefit. Options that should be chosen last of all – LL – are those that are not very viable and bring little environmental benefit.

Table 3.2 Foundations and retaining structures

Foundations and retaining structures	**Reuse in situ**		**Use of salvaged/ reconditioned products/ reclaimed materials**		**RCBPs**	
Chapter 4 **Element/component/product**	**Viability**	**Environmental benefit**	**Viability**	**Environmental benefit**	**Viability**	**Environmental benefit**
FOUNDATIONS						
Pad/strip foundations	H	H	–	–	H	H
PILES						
Timber piles	M	H	L	L	–	–
In situ concrete piles	H	H	–	–	M	H
Driven concrete piles	H	H	L	L	–	–
Driven steel piles	H	H	M	M	–	–
Steel sheet piles	H	H	H	H	–	–
Tubular steel piles	H	H	M	H	–	–
Screw piles	H	H	H	M	–	–
MISCELLANEOUS						
Soil stabilization	M	H	–	–	–	–
Fill	M	H	M	H	H	H
Embedded retaining walls	M	H	–	–	M	M
Gabions	M	H	M	H	H	H
Crib walls	M	H	M	H	M	M
Geotextiles	M	M	L	L	M	M
Bentonite	M	M	M	L	–	–

Reuse in situ

In principle there is little difficulty in using estate agents to find an existing building with an intention that it be wholly or partially reused. There may be greater difficulty in finding a suitable building in the right place.

More difficult still may be finding a building that will be suitable for the new intended use. Expert advice will be needed *before purchase* as to whether proposed modifications are likely to be possible, for example turning a church into a number of residential units. Generally advice will be needed from an architect, a structural engineer, a services/ utilities engineer and possibly someone with an understanding of local heritage issues.

Table 3.3 Building structure

Building structure	**Reuse in situ**		**Use of salvaged/ reconditioned products/ reclaimed materials**		**RCBPs**	
Chapter 5 **Element/component/product**	**Viability**	**Environmental benefit**	**Viability**	**Environmental benefit**	**Viability**	**Environmental benefit**
MASONRY						
Brick walling with lime mortar	H	H	H	M	–	–
Brick walling with cement mortar	H	H	L	L	–	–
Concrete block walling (cement mortar)	H	H	L	L	H	H
Glass block walling	H	H	M	L	M	M
Natural stone rubble walling	H	H	M	M	–	–
Natural stone ashlar walling/dressings	H	H	H	H	–	–
Cast stone ashlar walling/dressings	H	H	M	H	H	H
Precast concrete sills, lintels, copings and features	H	H	M	M	H	H
STRUCTURAL TIMBER						
Timber framing (traditional)	H	H	H	H	–	–
Timber floor beams	H	H	H	H	–	–
Timber roof trusses (modern)	H	H	L	M	L	L
Glulam timber members	H	H	H	H	L	L
Timber panels (chipboard etc.)	M	M	L	M	M	M
STRUCTURAL IRON AND STEEL						
Cast-iron columns and beams (mainly pre-1870s)	H	H	M	L	–	–
Wrought-iron columns and beams (mainly 1850s–1880s)	H	H	L	L	–	–
Wrought-iron roof trusses (mainly pre-1880s)	H	H	L	L	–	–
Structural steel framing (post-1880s)	H	H	H	H		–
Steel roof trusses (post-1880s)	H	H	M	H		–

Table 3.3 Building structure (continued)

Building structure	Reuse in situ		Use of salvaged/ reconditioned products/ reclaimed materials		RCBPs	
Chapter 5 Element/component/product	Viability	Environmental benefit	Viability	Environmental benefit	Viability	Environmental benefit
IN SITU CONCRETE/PCC						
In situ reinforced concrete frame	H	H	L	L	L	M
Formwork for in situ concrete	–	–	H	M	M	M
Precast concrete frame structures	H	H	M	M	–	–
Precast concrete blocks/planks	H	H	L	M	M	M
FLOOR STRUCTURES						
Brick jack arches	H	H	L	L	–	–
Precast concrete/ceramic blocks	H	H	M	M	M	M
Precast/composite concrete decking	H	H	M	L	M	M
Precast concrete planks	H	H	M	L	–	–
Profiled sheet metal decking	H	H	L	L	–	–

Use of salvaged or reconditioned products and reclaimed materials

There are three primary sources of goods and materials that can be reclaimed from one location and reused/recycled in another:

1 buildings that are about to be demolished;
2 dealers in various salvaged goods and materials;
3 specialists in reconditioned equipment.

Buildings due to be demolished

The most convenient opportunity for salvaging goods and materials from a building being demolished is on the site of a proposed new building. Even in the case of a building that is being demolished with no concern for reuse or reclamation of goods and materials, it may be possible to use some of the masonry rubble on the site for landscaping or fill. Otherwise, the most obvious way of finding buildings due for imminent demolition is through demolition contractors who have been approached by the owners of buildings.

Modern demolition is usually highly mechanized and requires expensive plant and equipment. Consequently, there are strong commercial pressures to ensure full utilization and demolish buildings as quickly as possible. While demolition contractors can easily be found through trade associations, the internet or phone directories, it is important to consider their specialization. A growing number of contractors, sometimes called 'reclamation contractors' are now specializing in the careful dismantling of buildings using more traditional, labour-intensive demolition methods that maximize the quantities of goods and materials that are salvaged for the reclamation market. Prior to any demolition an audit should be undertaken to

Table 3.4 Building envelope

Building structure	Reuse in situ		Use of salvaged/ reconditioned products/ reclaimed materials		RCBPs	
Chapter 6 **Element/component/product**	**Viability**	**Environmental benefit**	**Viability**	**Environmental benefit**	**Viability**	**Environmental benefit**
CLADDING & COVERINGS						
Timber board cladding	M	M	M	H	L	L
Sheet/board cladding	M	M	L	M	L	L
Profiled metal sheet cladding	M	M	H	H	–	–
Stone panel cladding	H	H	H	H	–	–
Concrete slab cladding	M	H	L	H	L	M
Glazed façades	M	H	L	M	–	–
Slate/tile cladding/roofing	H	H	H	H	M	M
Malleable sheet coverings/cladding	M	M	M	L	–	–
Thatch	L	L	–	–	–	–
WATERPROOFING						
Cementitious coatings	H	M	–	–	–	–
Asphalt coatings	H	M	–	–	H	H
Liquid applied coatings	H	L	–	–	–	–
Roof membranes	M	L	L	L	H	H
Felt/flexible sheets	M	L	L	L	M	M
Windows						

establish the equipment and other valuable items that can be removed in the soft strip and the approximate quantities of other items and materials with potential for salvage and reclamation (see Appendix A).

Buildings due for imminent demolition can be found in a number of other ways, including:

- Some architectural salvage firms advertise forthcoming demolitions on their websites, especially those where it is planned to salvage goods and materials.
- Main contractors or project management firms are often aware of the future demolition plans of some of their clients. They may also be aware of forthcoming demolitions from within their firms and will also be aware of many other projects, some of which are likely to involve a building being demolished.
- Through the local planning authorities who are likely to know of buildings on sites for which planning applications have been made for a new building.
- Major property owners will usually have a plan for their estate that will include plans for some buildings to be demolished.

Table 3.5 Enclosure, interiors and external works

Enclosure, interiors and external works	**Reuse in situ**		**Use of salvaged/ reconditioned products/ reclaimed materials**		**RCBPs**	
Chapter 7 **Element/component/product**	**Viability**	**Environmental benefit**	**Viability**	**Environmental benefit**	**Viability**	**Environmental benefit**
SPACE ENCLOSURE						
Rigid sheet construction (wood-based)	M	M	M	M	M	L
Plasterboard construction	H	M	L	M	M	M
Board/strip construction	M	M	L	L	–	–
Panel/slab partitions	M	M	L	L	M	M
Raised floors	H	H	H	H	–	–
Suspended ceilings (tiles)	H	M	M	M	M	M
Acoustic panels/tiling	H	M	M	M	M	M
Thermal insulation	H	H	L	L	H	H
Noise-proofing/insulation	H	H	L	L	H	H
WINDOWS etc.						
Timber windows/rooflights (glazed)	M	H	H	M	L	L
Steel/aluminium windows/ rooflights (glazed)	M	H	H	H	–	–
uPVC windows/rooflights (glazed)	M	H	L	L	–	–
Window glass (single sheet)	H	H	L	L	–	–
Sealed glazing units	H	H	M	L	–	–
Blinds	M	M	H	L	L	L
DOORS etc.						
Wooden doors (and wood fibre)	H	M	H	M	H	L
Glass doors	H	L	H	L	L	L
Screens/louvres (unglazed)	H	L	H	L	L	L
Shutters/hatches	H	L	M	L	L	L
STAIRS etc.						
Stairs/balustrades	H	M	H	M	L	L

Table 3.5 Enclosure, interiors and external works (continued)

Enclosure, interiors and external works	**Reuse in situ**		**Use of salvaged/ reconditioned products/ reclaimed materials**		**RCBPs**	
Chapter 7 **Element/component/product**	**Viability**	**Environmental benefit**	**Viability**	**Environmental benefit**	**Viability**	**Environmental benefit**
SURFACE FINISHES						
Screeds/trowelled flooring	M	M	–	–	–	–
Plastered coatings	M	M	–	–	–	–
Rigid tiles	M	M	L	H	M	M
Flexible sheet/tile coverings	L	L	L	L	M	M
Carpets (fitted)	L	L	M	M	M	M
Carpet tiles	M	L	M	M	M	M
Carpet underlay	L	L	L	M	M	M
Raised floors	H	M	H	H	–	–
Timber board	H	M	H	M	–	–
Wood block	H	M	H	M	–	–
Painting	H	L	–	–	L	L
FURNITURE AND EQUIPMENT						
Architectural ironmongery	H	L	H	H	–	–
Signage	M	L	L	L	H	M
Fireplaces	H	M	H	H	–	–
Furniture (not fitted)	H	M	H	H	M	M
Kitchen furniture	H	M	L	M	M	M
Kitchen equipment	H	M	H	H	–	–
SANITARY, LAUNDRY, CLEANING EQUIPMENT						
Sanitary, laundry & cleaning equipment	M	M	L	L	–	–
Toilets, sinks, baths	M	M	M	M	–	–

Table 3.5 Enclosure, interiors and external works (continued)

Enclosure, interiors and external works	Reuse in situ		Use of salvaged/ reconditioned products/ reclaimed materials		RCBPs	
Chapter 7 Element/component/product	Viability	Environmental benefit	Viability	Environmental benefit	Viability	Environmental benefit
PAVING, PLANTING, FENCING & SITE FURNITURE						
Stone setts, pavings, edgings etc.	H	H	H	H	–	–
Concrete pavings, edgings etc.	H	H	H	H	H	H
Fencing and barriers	H	M	M	M	H	H
Site/street furniture	H	M	M	M	M	M
Planting (pots etc.)	H	M	H	M	M	M
Garden/park furniture	H	M	M	M	H	H
Playground equipment	H	L	M	L	M	M

Table 3.6 Mechanical and electrical services

Mechanical and electrical services	Reuse in situ		Use of salvaged/ reconditioned products/ reclaimed materials		RCBPs	
Chapter 8 Element/component/product	Viability	Environmental benefit	Viability	Environmental benefit	Viability	Environmental benefit
MECHANICAL SERVICES						
HEATING, COOLING AND REFRIGERATION						
Heating boilers	L	M	M	H	–	–
Transformation and conversion of energy (chillers)	M	M	H	H	–	–
Measuring, detection and control devices	M	L	M	L	–	–
VENTILATION AND AIR-CONDITIONING						
Impelling equipment (fans)	M	M	H	H	–	–
Air treatment (humidifiers)	M	M	M	H	–	–
Distribution (ducting)	M	M	L	M	M	M
SUPPLY/STORAGE/DISTRIBUTION OF LIQUIDS/WASTE-HANDLING EQUIPMENT						
Water and general supply/storage/ distribution	H	H	M	M	–	–

Table 3.6 Mechanical and electrical services

Mechanical and electrical services	Reuse in situ		Use of salvaged/ reconditioned products/ reclaimed materials		RCBPs	
Chapter 8 **Element/component/product**	**Viability**	**Environmental benefit**	**Viability**	**Environmental benefit**	**Viability**	**Environmental benefit**
Steam supply/storage/distribution	M	M	L	L	–	–
Gas supply/storage	M	M	L	L	–	–
Gas distribution (pipes)	M	M	L	L	–	–
Liquid fuel supply/storage/distribution	M	L	L	L	–	–
Fixed fire suppression systems	M	L	L	L	–	–
Wet waste handling equipment	M	M	L	M	–	–
Solids waste handling equipment	M	M	L	M	–	–
Gaseous waste handling equipment	M	M	L	M	–	–
Drainage pipes	M	H	–	–	H	H
ELECTRICAL SERVICES						
ELECTRICAL POWER AND LIGHTING SERVICES/PRODUCTS						
Power storage devices	M	H	H	H	–	–
Transformation devices	M	H	H	H	–	–
Electric motors	M	H	H	H	–	–
Treatment devices	M	H	L	H	–	–
Distribution devices	M	M	L	M	–	–
Distribution cable	M	M	L	M	L	L
Terminal devices (sockets etc.)	M	M	M	M	M	M
Terminal devices (luminaires)	M	M	M	M	M	M
INFORMATION AND COMMUNICATION SERVICES/ PRODUCTS						
Safety and security information systems	L	L	L	L	–	–
Building management systems	L	L	L	L	–	–
Communication cables	L	L	L	L	–	–
TRANSPORT SERVICES/ PRODUCTS (Lifts etc.)						
Lifts	M	H	M	H	–	–
Escalators, conveyors	M	H	L	H	–	–
Electric motors	M	H	H	H	–	–

Salvage

Today there are two main ways of finding materials and goods that have been salvaged from building demolition for reuse or recycling: salvage yards and web-based information exchange.

It is the business of salvage yards to obtain goods and materials from buildings being dismantled or demolished and sell them on to others at a profit. They clearly incur costs in finding, obtaining, transporting, cleaning and storing these goods and materials. Their price is thus always greater than the basic cost of removing them from the demolition site. In many cases their price will also reflect their scarcity and architectural value. Most importantly, only those goods and materials will find their way to salvage yards that the dealers consider they will be able to sell with an adequate profit margin.

The following list indicates the range of goods and materials to be found at salvage yards:

- reclaimed bricks (especially ones dating from before 1900 when lime mortar was commonly used);
- roofing materials (tiles, slates, chimney pots);
- flooring materials (stone flags, quarry tiles, floor boards, wood-block flooring, stone setts);
- some structural timber and steelwork;
- timber and metal window frames and doors;
- staircases (cast iron and timber);
- stained glass windows;
- sinks, basins and baths;
- architectural ironmongery (door and window fittings, bathroom and kitchen fittings);
- radiators (mainly cast iron);
- fireplaces, grates and surrounds (marble, slate, cast iron);
- light fittings (from chandeliers to fluorescent light fittings);
- moveable partitions, shelving;
- miscellaneous furniture and fittings;
- garden ornaments and furniture (statuary, sundials, seats, fencing, gates).

Different dealers have different specialisms and the goods available will vary with location due both to local supply and demand and to the cost of land or property for storage. The premium market is found in the architectural salvage yards where goods and materials of high value can be obtained, usually a function of their scarcity or visual appeal. Some architectural salvage firms dealing in high-value goods circulate catalogues of what they have available (Kay, 2000; www.salvo.co.uk). Most salvage firms nowadays run websites that catalogue and illustrate the goods available and these have made considerably easier the task of locating salvaged goods and materials. Nevertheless it will usually be necessary to inspect goods closely before purchasing them since their quality will directly affect their suitability and their value.

In addition to salvage firms that often use their websites as an on-line catalogue, a number of organizations have set up a 'virtual' market place for reclaimed goods and materials. These work more like the 'wanted' and 'for sale' columns of a local newspaper. Advertisements are placed by the owners of goods and materials available for reuse or recycling and by people who have particular needs for reclaimed goods and materials. While many of these operate at a local level, there is also one national scheme – the Salvo Materials Information Exchange (www.salvomie.co.uk).

To the list of items likely to be found through salvage dealers, it is worth adding some key items that are generally not found this way, but which could be salvaged if they were identified prior to building demolition; in other words, if a demolition contractor or salvage firm knew there was a buyer for the goods prior to demolition. This would include:

- structural steelwork;
- windows and doors (including frames);
- cladding from the building envelope;
- raised floors and suspended ceilings;
- air distribution equipment (fans, filters, ducts etc.)
- various items of plant and building services equipment that might also be returned to firms for reconditioning (see below);
- distribution frames, racking for equipment and cables, trunking;
- lifts.

Providers of reconditioned plant and equipment

Salvage yards do not deal in working equipment. For this the intending buyer will need to approach firms that specialize in reconditioning. Often this can mean the firm that originally made the equipment or dealers who have set up in the business of servicing and reconditioning certain items of plant equipment. In essence this market is not dissimilar from that for second-hand cars and reconditioned parts for cars, such as engines and gearboxes. Generally this market place caters for medium/large items with high residual value.

The following list indicates the range of goods and equipment that can be obtained as reconditioned items, usually including a warranty of performance:

- diesel generators (usually stand-by generators);
- transformers;
- medium/large electric motors;
- switchgear, starters, circuit breakers;
- closed-circuit television (CCTV) and other security systems;
- heating boilers;
- chillers for air-conditioning;
- large fans and blowers;
- medium/large pumps;
- valves (especially motorized) and large-section pipes;
- storage tanks for various fluids;
- lifts;
- carpets and carpet tiles.

Nowadays the suppliers of these and other reconditional goods are easily found on the internet.

Recycled-content building products

There is now a growing market for recycled-content building products in the US, UK and many other countries. Apart from a normal search on the internet, a number of product directories are thriving that specialise in dealing only in RCBPs. As with all product directories, they are only as good as the material provided to them by suppliers, but the quality of this information is generally of a high standard.

The following list indicates the range of goods and equipment that can be obtained as RCBPs:

- precast concrete blocks, lintels etc. made using secondary (recycled) aggregate;
- reconstituted stone – precast concrete made using aggregate consisting of post-industrial, high-quality waste from stone working (marble, granite, etc.);
- roof tiles;
- a wide range of products made from recycled polymers, including pipes and ducts, guttering, downpipes and water butts, plastic lumber boards, kitchen furniture, garden furniture, playground equipment;
- cellulose fibre insulation made from processed waste paper;
- cardboard tubing made from processed waste paper;
- gypsum plasterboard;
- flooring and roofing tiles etc. made from recycled car tyres.

Ensuring reclamation, reuse and recycling happens

As we have seen in previous chapters, there are many reasons why reuse and recycling does not happen as much as it might. Underlying most of these are the two simple facts that little reward may accrue to the many and various players involved, and that the project must be organized and undertaken in ways that differ from 'normal' projects – human inertia is a highly effective barrier to change.

In the absence of a legal compulsion to reuse and recycle, the essential requirement is a client or developer who wants to do it or, at least, is not against the idea. This may require some tolerance to the duration of the design and construction phases since they will depend entirely on the supply of suitable materials and goods. For some elements of the building, it may also mean accepting that reused or recycled materials and goods may not be the cheapest option.

No less important is the need to engage a design team that is enthusiastic or willing to go down the reuse/recycle route, will accept the disruption to their normal working practices and is prepared to take the initiative when it comes to overcoming the many hurdles that are likely to present themselves.

Finally, it will be essential to employ contractors and sub-contractors who are committed to reuse/recycling. They will need to become involved in the project much earlier than usual, early in the design stage, in order to begin the process of sourcing suitable materials and goods. If not, an uncooperative contractor will easily be able to exert pressure on the rest of the project team to use new materials and goods. The two most common methods are simply to fail to find reused/recycled alternatives or to quote unreasonably high construction costs, using 'unfamiliar practices' as the excuse.

There are no secret ways of ensuring that reuse and recycling happen, any more than there are secret ways of ensuring that 'normal' projects are completed on time, within budget and free from defects. A reuse/recycling project simply has a number of different and additional issues to address. The key ones are these.

Overall project aim

As with all projects, the desired environmental performance will only be achieved if it is clearly identified and written down at the beginning of the project (Addis and Talbot, 2001). Reuse or recycling may be the main objective, or form just one part of a wider objective, for example:

- a building made only of reused components;
- a building made only of recycled materials;
- a building with minimum environmental impact (an assessment tool must be specified);
- a building with as many recycled/reused goods and materials as practical;
- a building that will achieve a Platinum LEED Rating (including appropriate reuse/recycling);
- a building with a recycled content of 10 per cent by value or 20 per cent by mass (for example);
- a building that achieves a certain score using an agreed reuse/recycling assessment method;
- a building that will gain all the reuse/recycling credits of LEED or BREEAM.

Having agreed an objective, it may be wise to agree a second-best objective to allow compromise in the event that the first objective proves to be too onerous (in cost, time). Only if the objective has been defined will it be possible to manage the design, procurement and construction processes to ensure the objective is achieved, and to verify that it has been achieved.

The design and construction team

It will be essential that the key members of the project team have been selected for their knowledge and understanding of reuse and recycling, and for their commitment to achieve it. These issues will need to be addressed when requesting expressions of interest, fee proposals and tenders. Firms and individuals will need to be assessed as to their understanding and experience of reuse and recycling when being interviewed.

Since it is unlikely that all members of the project team at all stages of the project will be adequately educated about reuse/recycling, it will be necessary to provide training for those who need it by means of workshops and guidance notes.

Specifying degrees of reuse/recycling

The particular type and amount of reuse/recycling will need to be specified in at least three ways at different stages of the project:

- in the project brief, for the building as a whole;
- during design development, element by element;
- in the specifications of materials and goods in employers' requirements.

At each stage it will be important to be able to assess/measure the type or proportion of recycled/reused goods and materials to ensure that what was required has been delivered (see above).

Design

In order that recycling is considered at the design stage, time will need to be devoted to the issue and this will add to the design costs, both directly and in the additional effort needed to evaluate the benefits of using reclaimed goods and materials against any additional costs.

Using reclaimed products and materials is sometimes perceived to increase the risk of failure to meet the required performance or durability of a building element. While such risks can be eliminated by testing and recertification, prejudices may still remain.

It will be particularly important that design team members become familiar with the world of reclaimed and recycled materials in order that they understand the range of opportunities open to them (for example Coventry and Guthrie, 1998; Coventry et al, 1999; IWMB, 2000; Kay, 2000; Kernan, 2002).

It is also believed by some people that building design codes encourage designers to use only new materials and products in preference to reusing reclaimed ones. While most design codes assume that new materials are normal practice, in fact most also include the opportunity to apply the code to reclaimed and recycled products, for example through the use of test data. Many countries produce their own industry guidance notes and codes for testing and reuse of a number of materials – brick, concrete, iron and steel structures. A client and design team need to understand that this apparent hurdle can easily be overcome in a technical sense, though it may incur costs and require additional time compared to using virgin products.

Design is always an iterative process, and designing with reuse and recycling in mind adds to the number of iterations. There will need to be a clear separation between the initial, outline scheme that defines the special aspects and overall appearance of the building, and the detailed design. The detailed design can only be undertaken when much of the equipment, components and materials has been found.

Specification

Specification clauses relating to reuse and recycling follow the same rules, so to speak, as the specification of all materials products and services in the building industry. There are now many useful sources of guidance for specifications relating to 'green' buildings and these usually include some mention of reclaimed goods and recycled materials (e.g. NBS, 1997; EPA, 2000; NES, 2001; NGS, 2004). Underlying all specification is the requirement that there must be ways of verifying that whatever was specified has actually been done. This is the primary purpose of standards for materials and design methods, agreed testing procedures, quality control measures, warranties and so on. Each country has its own set and these will be familiar to the building communities of those countries.

It is rare to find that reused or recycled products and materials are proscribed in standards. More likely, there may simply not be the test data available to provide the authoritative reassurance that designers and insurers need. For this reason, an essential part of the specification of reused and recycled goods and materials should include the sort of tests that will need to be done to verify adequate performance. In fact, most of these will be the same tests used for new materials and products. The main difference will be who does the testing and when. When reusing and recycling, the project team for a building need to take on some of these roles that are normally undertaken by suppliers of materials and products.

Cost plan

The cost plan for the building will need to be more adaptable than is usual for a 'normal' project. It will need to be able to respond to the uncertainties of the second-hand market place. It will also be necessary to take into account costs such as refurbishment and testing not usually encountered in buildings.

Project programme

Similarly, the project programme will need to be as flexible as possible to cater for delays while hard-to-find goods are located and to allow time for refurbishing and testing materials and goods. It is quite likely that delays will arise due to delays in other projects, for example the delayed demolition of a building identified as a source of reusable building materials or components.

The programme will be a valuable tool for communicating to the project team how the sequence of activities differs from a 'normal' project.

Finding materials/goods

It will be essential to find suitable materials and

goods prior to detailed design, and possibly earlier if the appearance of the building will be a major issue for the planning authority.

Some materials and goods will be found in salvage yards, others in buildings due for demolition and others from suppliers of reconditioned goods and RCBPs. It will be important to allocate clearly the responsibility for finding the various reused/ reclaimed components and materials for the building.

In the case of goods and materials salvaged from buildings due for demolition, it will first be necessary to undertake an audit of the building, and to assess the suitability of components and materials for reclamation, refurbishment and reuse. Both designers and the contractor will need to be involved in this process to ensure the products identified meet all the relevant criteria.

Once selected, arrangements will need to be made for the careful deconstruction of the elements identified and their transport to a site for storage prior to refurbishment. Details of the products and materials will need to be fed back to the design team so that the detailed design can be undertaken, and to the contractor so that a suitable method statement can be developed.

Demolition

For goods and materials found in buildings due for demolition, it will be necessary for the project team to work closely with the owner of the building and the demolition contractors (see Appendix A and ICE, 2004).

It may be that the demolition contractors have bought the building prior to demolition in order to maximize the money they can make from its disposal. In order to retain maximum control of the demolition process, and access to the fruits of deconstruction and demolition, it may well be worth the client for the reuse/recycling building buying the building due for demolition.

Prior to the heavy demolition machinery arriving at the site, the high-value goods will usually be removed in the soft strip. The reclamation may be procured in three different ways:

1. The demolition contractor undertakes the soft strip, reclamation of goods and materials and demolition of the building.
2. The demolition contractor employs a specialist 'reclamation contractor' to undertake the soft strip and reclamation of goods and materials, and the demolition contractor undertakes the demolition of the building and disposal of the remaining materials;
3. The demolition is led by a specialist reclamation contractor who undertakes the soft strip and reclamation of goods and materials, and employs a specialist demolition contractor to undertake the remaining demolition of the building and disposal of the materials.

Each method has its advantages and a careful assessment and cost comparison will need to be made to select the most appropriate in a given case. It will be mostly dependent on the building being demolished and the quantities and nature of the materials and goods likely to be reclaimed.

Invitations to tender for reclamation and demolition will need to specify carefully the goods and materials to be reclaimed and ideally, should try to ensure that the timing of the process suits the reuse/recycle project. It will need to be clear in the tender price what goods and materials are to be reclaimed and sold, and what costs will be incurred for material sent to landfill.

Depending on the local authority where the demolition is being undertaken, there may be financial incentives or waste credits available for reducing waste through reuse and recycling.

Selection of materials/goods

When selecting which of several alternative materials or goods to buy, it will be necessary to use a robust strategy for deciding between them, balancing cost, availability and environmental benefit.

When selecting reclaimed materials and goods it can be helpful to treat them like natural materials, such as timber or stone, which vary both in their appearance and their properties. When procuring a façade of a building (using new natural materials), for example, it is usual to apply a full range of selection criteria using visual samples, range samples, a full-size mock up, testing offsite and onsite and taking control samples when completing the works. Such an approach can be appropriate for reclaimed materials and goods.

An important feature of the selection process is that it needs to allow for compromise. It will be inevitable that some reused/recycled items sought for the building will be difficult or impossible to find, will be significantly more expensive, or may cause greater environmental impacts than their new equivalents.

Storage

In the event that materials and building components are identified in a building prior to demolition, it will be necessary to find cost-effective storage for materials and goods after their removal from the building. Indeed, two periods of storage may be needed – one between their purchase and refurbishment and one after refurbishment prior to when they are needed in construction.

The main contractor will need to be involved in this aspect of the project as they are most directly concerned about the type, quality and availability of the goods and materials, and how they will be used in construction. A large contractor, working concurrently on many projects, may be able to help not only in finding buildings to be demolished, but also vacant areas of sites where goods and materials can be stored.

Planning

The materials, components and, hence, the appearance of the building will not have been finalized at the stage when a planning application is usually made. It may therefore be necessary to submit an outline application, followed by a detailed application once materials and components have been found.

A growing number of planning authorities provide guidance on 'sustainable construction' that usually includes recommendations to reuse and recycle. While no prescribed targets are generally included, some authorities require 'sustainability checklists' to be submitted with planning applications, with a requirement for a minimum score; some points or credits are likely to be available for reuse or recycling.

Project management

It is one thing to design a building to incorporate reused goods or recycled materials; it is quite another to ensure that it is actually constructed that way. Designers' proposals and intentions are often changed between detail design and the completion of a building. Those who control the final specification of materials and equipment will have the final say, and they are likely to be driven by other goals such as capital cost and the construction programme.

The wishes of the client or the design team will be achieved only if an effective delivery system is implemented. In this respect the process mirrors that needed to achieve good quality in finished buildings, which is usually delivered by implementation of an effective quality management system that can be certified against ISO 9001 (Quality Assurance). The delivery of a building incorporating reused/recycled goods and materials can be considered as one aspect of the environmental performance of the construction process and therefore could be addressed using an environmental management system similar to those covered by ISO 14001.

Insurance

Insurance policies generally make it financially unfavourable to use reclaimed materials. It will usually help with reuse and recycling if insurance companies are brought into discussions with specifiers and educated about the nature of the risks involved and how they can be managed.

Health and safety regulations

A project involving reuse and recycling will be subject to the same requirements to meet health and safety regulations as a normal project. It will be important to incorporate the reuse/recycling objectives into specifications sent out to obtain tenders and health and safety plans for the project. It will be important to ensure that contractors are familiar with the manner in which progress towards achieving such goals will be assessed.

In conclusion

Each material, component and system in a building has its own series of issues that need to be addressed when considering the opportunities for reclamation and reuse, or for using recycled materials. These issues are addressed in the following chapters that are devoted to the individual elements of buildings.

4 Design Guidance: Foundations and Retaining Structures

Building foundations

The construction activities of our ancestors in all towns and cities have left two important legacies: above ground, the built environment that forms such an important aspect of our very culture; and, below ground, the remains of those buildings that history has judged not worthy of preservation. The foundations left in the ground after a building has been removed vary according to the type of soil and its water content, which affect the load-bearing characteristics of the ground, as well as the loading on the foundations, which usually reflects the height of the building. Building foundations come in two basic types.

Pad or strip foundations are triangular in cross-section and spread the load from a column or wall over a large area of soil to prevent penetration into the soil. They also provide a good base, at the right height, upon which a column or post, or a wall is constructed. They are used when the soil is relatively strong and the loads from above are relatively low. For low-rise buildings pad and strip foundations are usually made of mass concrete, often mixed with rubble. When loads were higher, or the soil weaker, the pad or strip would be made of dressed stone or strong bricks. An even higher load-bearing capacity could be achieved using a grillage consisting of several layers of beams made of timber or, since the 1870s, steel, laid one upon the other to form a pyramid. The voids between beams were filled with rubble or concrete. Many medieval cathedrals were built on timber grillages and many buildings up to 10 or 12 storeys high were built in the late 19th century on steel grillages in Chicago, where the soil is notoriously weak.

Where a greater load-bearing capacity is required, piled foundations are needed. From Roman times up to the 19th century these were made of timber, sometimes capped with a metal sheath to enable the pile to penetrate the soil more easily and without damage. From the late 19th century piles of reinforced concrete and steel were used. The piles were driven by percussive action until they almost ceased to penetrate further with each successive blow. The piles resist further penetration either because they meet rock or an impenetrable layer of soil (end-bearing piles), or because of the friction between the ground and the pile itself (friction piles). In older buildings, a number of piles would be driven close together and a grillage built over the end of the piles to form the platform on which the building could be constructed. Since the use of steel and reinforced-concrete piles from the late 19th century, a pile cap is constructed on top of just one, two or three piles upon which the column of a steel or reinforced concrete frame building is constructed. Nearly all modern large buildings are constructed on piled foundations.

Tension piles perform a duty similar to ground anchors and hold a building down rather than keep it up. They are used in two circumstances: in a tall building when wind loads tending to overturn the structure lift the piles on the windward side; and in buildings with deep basements that may effectively float if the groundwater level is at or near ground level.

The legacy of previous construction on a building site may, then, be a large number and volume of building foundations. The geotechnical engineer has three options – to avoid the existing foundations, to remove them or to reuse them.

Avoiding existing foundations

The usual approach by engineers faced with existing foundations was, and is, to construct new piled foundations in the gaps between existing foundations. This option is likely to constrain the places where new piles can be driven. Some existing foundations are likely to be around the perimeter of the building where new piles are needed. Existing foundations may, in turn, also have been located to avoid even earlier construction, including archeological remains, or underground obstacles such as tunnels, services ducts, streams and so on. In a congested site, former groundwork can almost be considered as a form of contamination!

Removing existing foundations

Sometimes it may be necessary to remove existing foundations. Not only can this option be expensive – between two and five times the cost of a new pile – it will cause disruption to the geotechnical status quo, which may adversely affect the performance of other foundations left in the ground. As a general rule, soil is best left as you find it because it has probably arrived in its present condition and state of equilibrium over a period of decades or even centuries. Disturbing soil by removing foundations can change the flow pattern of groundwater under a building that may affect the strength of the soil; it may also release compressive stresses in the soil that are providing the friction necessary for the working of an adjacent friction pile; and it may reduce the horizontal soil pressures on foundations of an adjacent building and allow those foundations to move laterally, causing the building above to deform or crack.

Reusing existing foundations

The reuse of existing foundations will eliminate the cost and time needed to construct new foundations. However, the reuse of existing foundations has its own costs and also takes time to arrange and undertake.

It has long been common practice to reuse existing pad and strip foundations whenever possible. This avoids the costs of excavation and disposal of the excavated materials and saves on the costs of new construction. Generally, however, this can only be done when it is considered 'obvious' that the existing foundations are strong enough to carry the new construction. They must also be in the right location.

Where it is not obvious that the existing foundations will be adequate to carry the new loads, their reuse requires a reliable assessment of their strength. The traditional approach was to prove the existing foundations by test-loading them. If they failed, however, the task of replacement could be both costly and time-consuming.

Proving the capacity and suitability of existing foundations without test-loading has become technically possible during the 20th century as geotechnical engineers have come to understand the behaviour of soils more and more fully. Nevertheless, it is not yet common practice to reuse foundations for new buildings. The main barrier has been the inability to persuade developers, their funders and building insurers of the reliability of assessing the condition and capacity of existing foundations. In the last few decades this situation has begun to change as reliable assessment techniques have been developed and proven. The option to reuse existing foundations should nowadays always be considered (St John et al, 2000; CIRIA, 2006).

Finally, if new foundations are needed, it may be possible to use materials with a recycled content,

Table 4.1 Relative impacts of options for foundations in new buildings

	Reuse existing foundations	Use recycled materials in new foundations	Install new foundations	Remove existing foundations and replace with new foundations
Amount of carbon embodied in materials used	Low	Low	High	High
Amount of carbon consumed during installation	Low	High	High	Very high
Amount of carbon consumed during operation	Low	Low	Low	Low

Source: Adapted from Chapman et al, 2001

for example, concrete made with aggregate substitutes or cement replacement materials.

The relative environmental impact of each of these options can be evaluated using 'carbon accountancy', which assesses the amount of carbon (in the form of fossil fuels) needed to create the product. The table opposite shows the relative ranking for foundation strategies according to the amounts of carbon needed.

Reuse of foundations in situ

In order to ascertain the viability of reusing existing foundations, whether retaining walls, pad, strip, raft or piled foundations, a number of issues must be addressed in addition to those relating to the design of new construction:

- impact on surrounding buildings and their foundations;
- arrangement of existing foundations and the needs of the new building;
- acceptance of reuse by designers, clients, insurers and regulatory authorities.

Assessment of existing foundations

In all but the smallest buildings, there will need to be a formal process of establishing the capacity, integrity and durability of existing foundations. The precise function and the performance of existing foundations in their previous use will provide the first guide as to their likely suitability for any future reuse – foundations that have successfully supported a six-storey building for 100 years are likely to be able to do so for some time to come. This argument, however, is not sufficient on its own. A rigorous process must be followed by geotechnical engineers to assess the capacity, integrity and durability of the foundations.

1. A desktop study of archival information to determine:
   - previous uses of the site as far back into history as possible;
   - the as-built design of the existing (former) building and its foundations;
   - original design loadings and design life.
2. A survey of the existing building to confirm:
   - the as-built design of the existing (former) building and its foundations;
   - construction materials and methods used;
   - evidence of foundation performance – movement, settlement, and so on.
3. Examination of existing foundations/piles (at least the pile heads) to determine:
   - dimensions, construction and location of piles;
   - condition of existing foundations – corrosion, rot, condition and integrity of concrete, exposure of reinforcement;
   - taking/testing samples of piles to determine material characteristics and properties including strength and stiffness in compression and tension.
4. A ground investigation to determine:
   - stratigraphy of soil (borehole);
   - strength, stiffness and other geotechnical properties of the soil – pore water pressures;
   - contamination in soil and groundwater (sulphates, etc.) – chemical aggressiveness of the soil and groundwater.
5. In situ tests on and observation of piles to determine:
   - results of load-tests to establish capacity of a pile;
   - integrity of concrete in pile (dynamic, non-destructive testing);
   - behaviour of piles and surrounding ground on loading and unloading.
6. Design calculations, using the data gathered, to determine:
   - loads carried by piles in previous use (load take-down calculation for former building);
   - predicted load capacity of piles;
   - predicted settlement of piles;
   - predicted durability of piles.

The assessment process is summarized in Figure 4.1.

A minimum condition is that the capacity of the existing piles is greater than loads they are anticipated to carry in their new role. Special attention must be given to any change of loading on the piles, for example due to change of building use, and changes in the ground conditions since the foundations were first constructed (e.g. deterioration

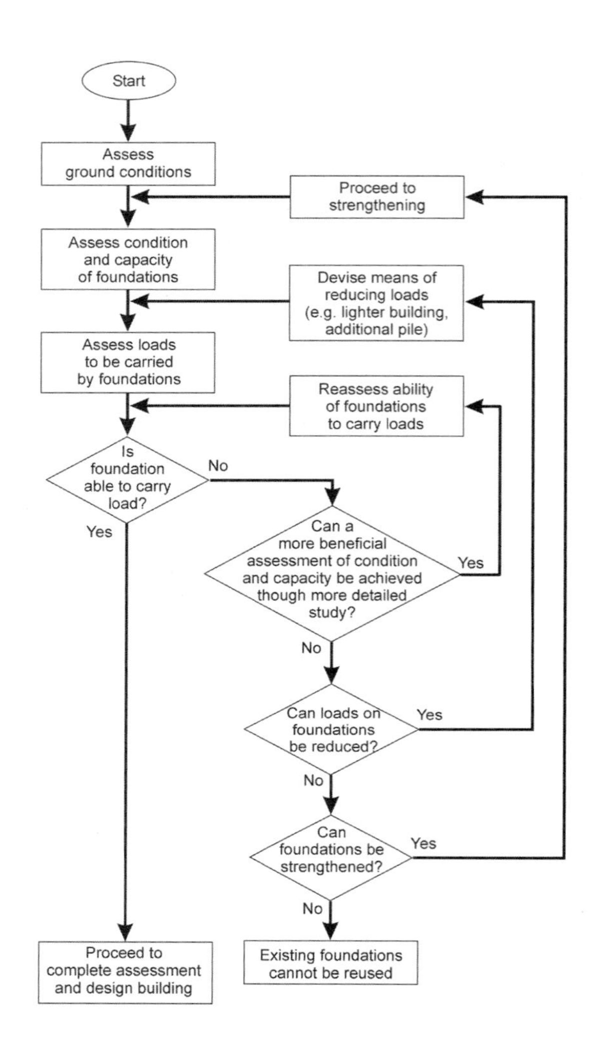

Figure 4.1 Flow chart of appraisal/assessment process

of materials, change of groundwater level, ground movement or settlement).

In general the condition, integrity and capacity of existing foundations are easier to establish if the building they formerly supported has been removed. If this is the case, however, it will be important to gather reliable data about the former building in order to be able to assess the load that the foundations actually carried just prior to their exposure. It is worth noting that foundations will usually moves upwards when the load from the building above is removed. If it is possible to measure this movement, it will provide valuable information about the properties of the soil. The removal of the building will also change the soil pressures beneath, which may change the apparent strength of the soil and its permeability to water.

First it will be necessary to undertake a load take-down to ascertain the loads that the foundations were formerly carrying. Drawings and design calculations for the building previously on the site will provide valuable information about the loads the piles were intended to carry. Ground investigation reports from the construction stage of the building may also exist. Such information, however, should not be relied upon without some checking – the building may not have been constructed as designed and there may have been alterations to the building and its foundations during its life.

Having established which foundations can be reused, their capacity and precisely where they are, the new building can be designed to suit the existing foundations. Three options are generally available:

1 Choose a column grid that coincides with the layout of the existing foundations.
2 Place a transfer structure between two or more piles that will support a new column.
3 Provide a number of new piles to augment the existing foundations, for example to accommodate a transfer structure between old and new foundations.

In general, a degree of flexibility will be needed on the part of the design team to achieve a workable solution.

When using an existing foundation, it will be important to ensure the wall or column loads are applied in precisely the same position as the former loading, otherwise the foundation may tilt when loaded or an eccentrically loaded pile will be subject to bending loads it was not designed to resist. When a mix of old and new foundations is used, a means must be found for catering for any differential settlement between the new foundations and the old ones.

If it is found that some foundations are damaged or have deteriorated, it may be possible to refurbish them or to take alternative measures to upgrade their capacity or durability by underpinning, grout injection to stiffen the surrounding soil, or sharing some loads with new foundations using transfer structures. A full range of suitable techniques can usually be found in books on building conservation (for example CIRIA, 1994; Beckmann, 1995).

To minimize the need for refurbishment, it will be important to ensure that ground products are not damaged during demolition, for example by ensuring construction plant is kept a safe distance from pile caps and retaining walls, or by providing temporary support or stabilization to prevent any damage to or movement of foundations exposed during demolition or excavation.

Where existing foundations are being reused, building insurers may be unwilling to insure the new building, especially as the previous owners, designers or contractors are unlikely to be willing to accept responsibility and liability for the existing foundations. A solution to this problem is for the designers of the new building, who have also assessed the existing foundations for reuse, to take responsibility for the existing foundations. Another is for the building owner to insure the building separately from the foundations. In either case, the cost of such insurance must be taken into account when considering the possibility of reusing foundations.

Pad, strip and raft foundations

Pad, strip and raft foundations can all be reused as long as there is no evidence of inadequate performance, and they are not required to carry a different type or larger load than in their former life. A geotechnical engineer will be needed to advise on the suitability of existing foundations for reuse. The formal procedure described in Figure 4.1 above must be followed.

The most common reason why pad, strip and raft foundations may not be suitable for reuse is the performance of the soil rather than the foundations themselves. All foundations settle under the action of the loads and the response of the soil beneath. Foundations may also move if the groundwater level changes and if tree roots grow near to or beneath a building. The usual consequences of such movements are cracks in the brittle elements of the building above – brickwork, plaster and rendering, glass and concrete.

Nevertheless, foundations that have moved may still be able to reused, for example by strengthening them or by adding additional foundations. A range of appropriate techniques can be found in books on building conservation (for example CIRIA, 1994; Beckmann, 1995).

Table 4.2 Methods of assessing existing pile capacity

Method	Description
Original design calculations	If available, the original design calculations will provide justification of the situation as perceived at the time; however, they may not be compatible with modern design codes or current geotechnical theories. They may also contain errors and the ground conditions may have changed (e.g. change in groundwater level)
New calculations	With knowledge of the soil conditions and pile dimensions, calculation of pile capacity can be undertaken using current theory and design methods
Pile testing	In situ tests can be carried out on piles. Static load-tests can use the original building as reaction; every tension pile will need to be tested to its full capacity. Ideally, all compression piles will be subject to testing. In practice this will usually be too costly and only a sample of the piles will be tested. Statistical analysis of results from a relatively small sample – maybe only 1–2 per cent – can usually provide a sufficient level of confidence to make predictions about all the foundations on a site
Load take-down	The load actually carried by existing piles can be confirmed by undertaking load take-down calculations based on the design of the structure of the original building. As with the assessment of all existing structures, good engineering judgement is needed in choosing an appropriate factor of safety or load factor when assessing potential for reuse

Source: adapted from Chow et al, 2002

Piled foundations

The consideration to reuse piled foundations has arisen from two situations that are increasingly common in inner-city sites: either existing foundations exposed after building demolition appear to be in good condition and are likely to be suitable for carrying the loads imposed by a new building, or a site is congested with the foundations of two or even more buildings previously occupying the site (St John et al, 2000; Chapman et al, 2001; Coles et al, 2001; Chow et al, 2002).

The viability of reusing piles is assessed by following the general procedure described above. Whatever materials the piles in foundations are made of, their capacity may be assessed using the approach outlined in Table 4.2.

Reinforced concrete is now the most commonly used material for piles and can be used to form a number of different types of piles, including driven precast, reinforced or prestressed concrete piles, and both driven or bored cast-in-place concrete piles. All these types are likely to be found in buildings constructed within the last half-century or so. In order to determine the durability of the concrete, that is, its remaining life, it will be necessary to take core samples of the concrete to determine the petrography of the concrete aggregate, any chemical deterioration of the concrete, and also to assess the aggressiveness of the chemical environment in the ground (BRE, 2003).

Steel piles are likely to have suffered some corrosion, depending on the presence of water and the chemical nature of any contamination. It will usually be straightforward to evaluate this by visual inspection of the exposed parts of the pile and by chemical analysis of the soil and groundwater.

Timber piles were in common use until the mid-19th century and will be found supporting most buildings of historical, cultural and heritage interest. The serviceability of timber piles is demonstrated by the large number of such buildings that survive. The reason for the longevity of timber piling, contrary to our normal experience that timber is rotted by water, is that rotting occurs only in the presence of oxygen; underground and underwater, in the absence of oxygen, timber maintains its integrity and strength almost indefinitely. It will be possible to reuse timber piles as long as the assessment procedure outline above is followed (see the Tobacco Dock case study, below).

Ground retention and other products

A wide variety of structures are used to retain soil and they can all be reused in situ, subject to an engineering analysis of the loads they must carry, their design and construction and their condition.

Reinforced concrete retaining walls may be made of precast units or cast in situ. Their primary purpose is to retain soil that would otherwise fall away from an exposed vertical soil face. Three main types are used: vertical retaining walls with their foot embedded in the ground; walls held in place using ground anchors that tie the wall back into the earth being supported; and L- or inverted-T-shaped precast concrete units that rely partly on their dead weight and geometry to retain the soil. An essential part of designing all three types of retaining wall is to prevent the failure of the entire wall together with a large body of the soil it is retaining by what is known as a slip-circle failure. To predict the conditions of such a failure it is essential to know the geotechnical properties of the soil and, especially, the presence of groundwater. It may be necessary to provide drainage to prevent water collecting in the retained soil and precipitating a catastrophic failure. When a retaining wall is assessed for reuse it is especially important to ensure this failure mode cannot occur. It is also important to ensure the weight of soil retained is kept the same and no additional loads must be placed on the retained section of soil as this could cause failure. This caution must be heeded during the construction process as well as in the final state.

Crib walls consist of reinforced concrete box frames, stacked one upon the other, and filled with soil. They are usually raked, for example to form the exposed retaining wall of a cutting. When assessing the possibility of reusing crib walls it is important to ensure that:

- individual members are not broken or fractured;
- all members are present;
- all connectors are in good working order;
- the loading on the crib wall has not changed.

Gabions consist of cobbles, stones or demolition rubble retained in a basket made of steel mesh to form large blocks. They rely on their mass to retain soil and so their reuse depends on the condition of the steel mesh. However, it is likely that corroded steel mesh will be at the back of a gabion wall and hidden from view, making inspection difficult. Unfortunately, they can be an ideal breeding ground for rats, which may discourage their reuse.

Many techniques are available for improving the load-bearing capacity of soils and, in principle, any of these can be reused in situ, as long as it is possible for a geotechnics engineer to assess their capacity and their condition. If found to be in an inadequate state of repair, such groundwork can often be refurbished or repaired in situ. This option will frequently be cheaper and involve less disruption than new construction.

The load-bearing properties of soil can be improved by a number of *ground modification techniques* that can be used to increase the shear strength and stiffness. The most common techniques include:

- injecting cement or chemical grouting under pressure to introduce compressive pre-stressing;
- consolidation by vibrating;
- soil nailing;
- ground anchors;
- reinforcing the earth using a mesh of steel or a polymer geotextile (usually polypropylene).

Soil modified in these ways can only be reused if left undisturbed, which makes a normal soil investigation impractical. A pragmatic approach to reuse must be taken. Ground modified by soil nailing, for example, could be assessed for possible reuse by taking the following approach:

- Make assumptions about the integrity of the soil nails, based on soil conditions and previous performance.
- Make assumptions about the loading capacity of the soil nails from the as-built information.
- Supplement the existing soil nails with new nails to ensure an appropriate safety margin.

Ground anchors will have been designed to carry certain specified loads. Original design calculations will help ascertain the potential capacity for reuse, as long as it is judged that the anchors have been well constructed. They can easily be tested, perhaps at 150 per cent of their original design load, and then used at 90 per cent of their original design load to provide an added degree of confidence. With such testing they could be given a ten-year warranty.

Use of reclaimed products and materials

Due to the nature of ground treatment and ground retention products, most ground products are not suited to removal and reuse in a new location, as the original design would be based on ground conditions and it is not likely that identical conditions will exist on another site. In addition, it would not be possible to remove many of the different types of ground products without damaging them, thereby rendering them unfit for reuse once removed.

Piled foundations

As discussed above, there are good reasons for not removing existing piles and so it is unlikely that there will be opportunities to reuse piles that have been removed. This said, it would generally be possible to remove a driven steel pile with relatively little disturbance to the ground and adjacent piles. Such a pile could be considered for reuse in a different location as long as its condition was judged to provide the necessary durability.

The case of circular steel piles is a good example of the building industry being able to make use of a waste or redundant product from another industry. The UK firm Greenpiles reclaims surplus steel pipe from the oil industry and uses it to make circular steel piles for building foundations (Figure 4.2). The company inspects the steel pipes to ensure they have no significant damage or corrosion and offers the same warranty on their use as if the piles were new.

Screw piles also provide an opportunity for reusing piles. After a building has been demolished, these can be easily withdrawn leaving the ground as it was before the building was constructed. The piles can be inspected and reused if they are considered of adequate quality (Figure 4.3).

Figure 4.2 Reclaimed steel pipes are refurbished by the firm Greenpiles and used to form piles for foundations

Source: Greenpiles (www.greenpiles.co.uk)

Ground retention and other products

Interlocking steel *sheet-pile walls* are often used for temporary works, to retain both soil and water. Steel soldier piles (H-section) are also used in the construction of temporary retaining walls for trenches or single-sided excavation. Contractors regularly remove both types of pile for reuse many times, until they have been damaged too much by the driving or removal processes to fulfil their function.

Crib walls can be dismantled and reused in new locations. It is important to ensure that:

- individual members are not broken or fractured;
- all connectors are in good working order;
- the cribs are used in accordance with the manufacturer's guidance.

Gabions can easily be reused, provided that the mesh is in good condition and is not likely to fail during its proposed life. Where the mesh is not in good condition, it is possible to reuse the cobbles and replace the old mesh with new mesh. The note above about rats should be heeded.

Recycled-content building products

Foundations

The only significant opportunity for using recycled-content material in the construction of new foundations is the use of recycled concrete aggregate for in situ concrete work.

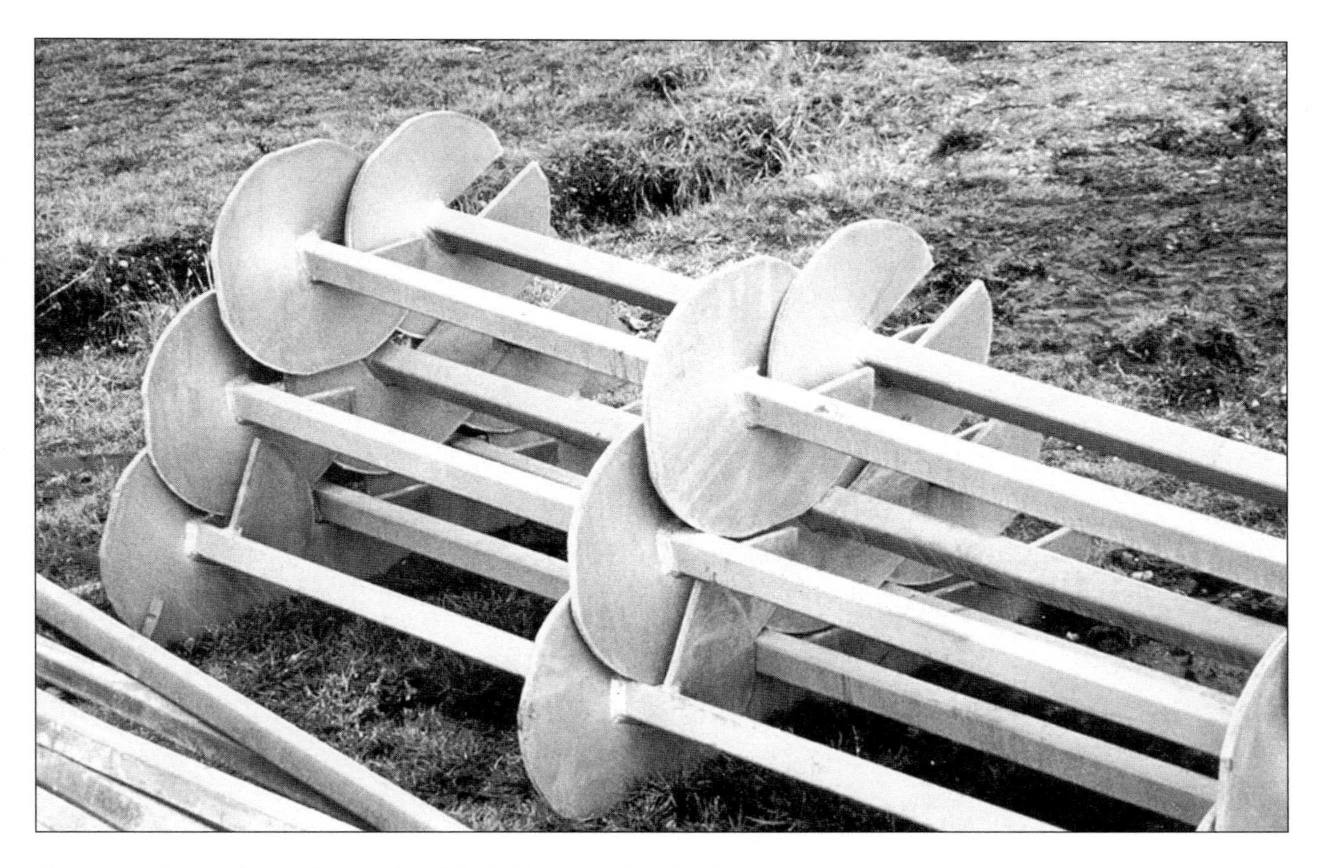

Figure 4.3 Screw piles used at Eastbrook End visitor centre, England

Source: Buro Happold
Note: Engineers, Buro Happold; architects, Penoyre and Prasad

Ground retention and other products

The main opportunity for using recycled-content material in the construction of *retaining walls* of both in situ concrete and *crib walls* made of precast concrete is the use of RCA for in situ concrete work.

Depending on the aesthetic requirements of a *gabion wall*, it may be possible to use recycled materials, especially demolition rubble, rather than new cobbles.

Used car tyres can be used to make low crib walls.

Geotextiles used in various techniques for ground improvement are available, made with a proportion of recycled-content polymers.

The most plentiful recycled-content material on most construction sites is the rubble arising from the demolition of concrete and brickwork. This can normally be used for a number of purposes that remove the need to import new materials to the site:

- filling unwanted holes on the site and making up the ground level prior to new construction or landscaping;
- backfilling behind cast-in-situ walls of concrete basements and such like;
- sub-base for access roads and car parks;
- piling mats – temporary bases for construction plant and machinery during auguring or pile-driving processes.

Figure 4.4 Tobacco Dock, East London, UK (1812–13): The 17-metre-span (54 feet) timber roof trusses are supported on cast-iron columns resting on granite vaults resting on timber piles

Source: Ove Arup and Partners

Case studies

Tobacco Dock, London

The original bonded warehouse was built in 1812–13 when London Docks was expanding to handle more and more imports of raw materials from around the world. The building has two storeys: the basement, consisting of masonry vaults on an 18 foot by 18 foot grid, was used to store barrels of wine; the upper storey consisted of an enormous roof with timber trusses spanning 54 feet at 9 foot centres supported on cast iron columns at 18 foot centres. The total area of floor covered was around 100 metres by 100 metres. The building was redeveloped in the 1980s to create a large commercial centre including shops, restaurants and cafes.

The refurbishment of the roof over the upper storey represented a major example of the reuse of an existing structure. A number of the timbers in the roof trusses had suffered decay at their supports and they were repaired with new timber connected to the old timbers using scarf joints. Some areas of the original roof had collapsed and it was planned to retain only around 85 per cent of the covered area. This effectively released a number of components of the roof structure to be dismantled and used to replace faulty components in other areas (Figure 4.4). It also meant that a number of the basement masonry vaults and the supporting piles beneath them were surplus to requirements. Samples of the masonry (brick and granite) were removed and tested to establish their strengths and stiffnesses. This was invaluable in enabling the structural engineers to confirm that the vaults would be able to carry the loads due to be imposed in the new use.

The foundations supporting each of the 100 or so granite columns of the masonry vault consisted of a brick pier built upon a platform of timber planks supported by eight timber piles. Each pile made of ordinary pine was 3 × 3 inches (225 × 225mm) square and around 12 feet (3.6m) long.

Modern building by-laws prohibit the use of timber in building foundations and the local district surveyor voiced his disapproval of timber piles. He indicated he would only give permission for the redevelopment if new concrete foundations were inserted beneath the brick piers of each foundation. Initially the client too was concerned about the possible reaction of the funders and future tenants of the development who might be apprehensive of a building with timber piles, especially in the light of the district surveyor's view. While underpinning every column was technically possible, and in fact done beneath a small area of the building, it would also have been a significant financial burden on the project.

At the time, in the early 1980s, there were still few precedents for reusing old industrial structures, either the foundations or the above-ground structure. It was the norm to demolish such buildings. Nevertheless, it was increasingly being realized that many such buildings were still in use with no apparent distress, and the importance of such buildings to the nation's heritage was also gradually becoming more widely recognized. Building conservation was beginning to embrace rather more than superficial repairs to old masonry, brickwork and windows.

Timber piles have been in constant use since Roman times and most of the world's medieval cathedrals and masonry bridges are built upon them. The key to timber's continuing preservation in foundations is that it must remain submerged in water. It is contact with oxygen in the air that causes timber to rot. A number of the piles at Tobacco Dock were investigated in situ by digging trial pits and they were found to be in good condition (Figure 4.5). Load-tests were carried out on three piles after which they were extracted from the ground and subjected to more detailed investigation and tests on the condition and properties of the timber. The piles were found to be capable of carrying the self-weight and the live loads anticipated in the refurbished building.

Figure 4.5 Several timber piles that would not be needed were extracted, inspected and found to be in excellent condition

Source: Ove Arup and Partners

The data collected were finally presented to the district surveyor and client, who were reassured that the original timber piles would be satisfactory. The only remedial action needed for the foundations was to ensure that the piles would always remain submerged in water in the future. Records of water levels showed that they were static; however, as a precaution, a system was devised that would maintain the current water level in case the situation changes in the future (Figure 4.6).

Location: London docks
Date: Original building 1812–1813; refurbishment 1985–88
Client: Tobacco Dock Development Ltd
Geotechical and structural engineers: Ove Arup and Partners
Architect: Terry Farrell Partnership
Refs: Courtney and Matthews, 1988; Mitchell et al, 1999

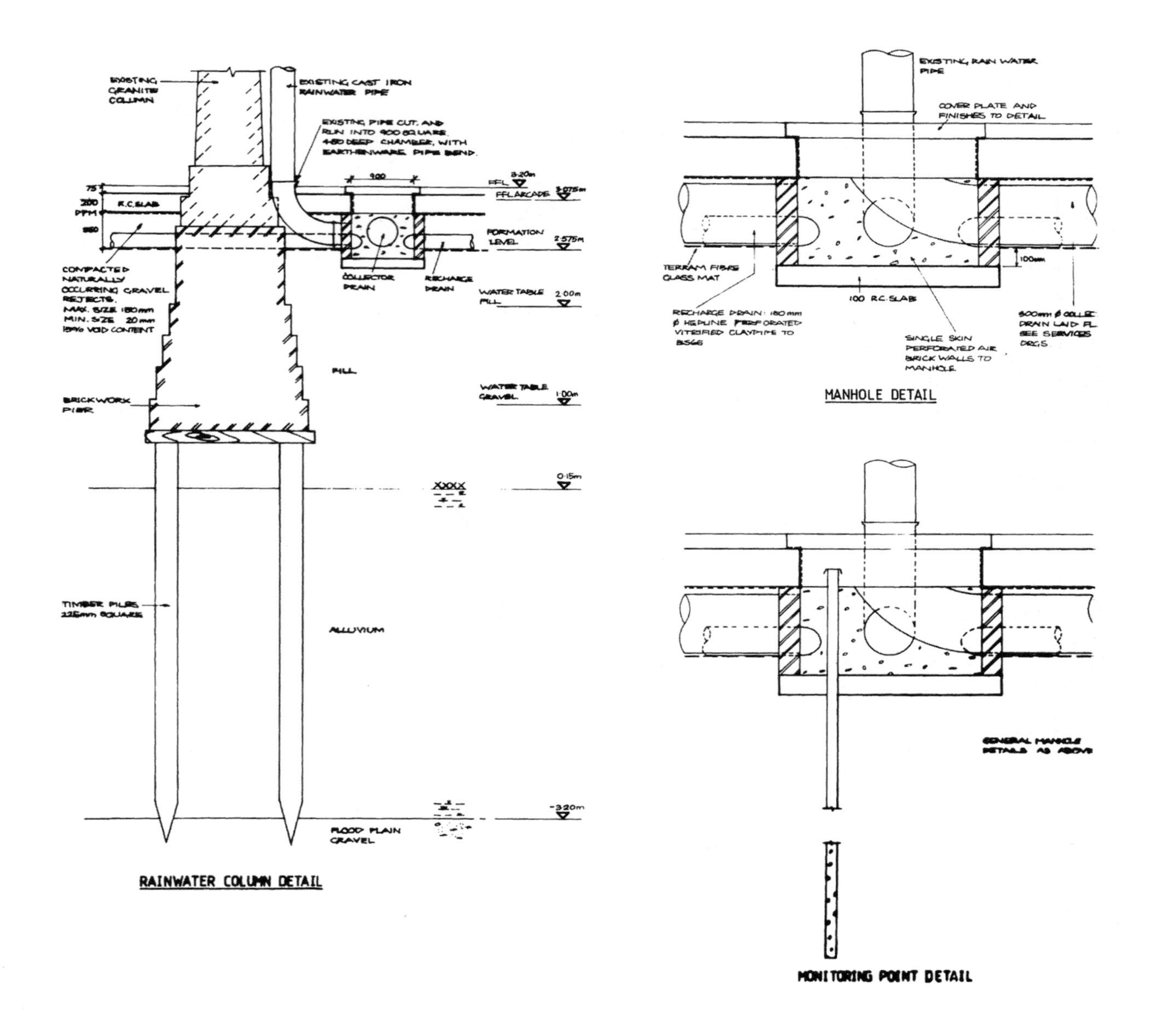

Figure 4.6 Details of the foundation and water recharge system to ensure the timber piles will never dry out

Source: Ove Arup and Partners

The Dearborn Center, Chicago

Figure 4.7 The tops of Jenney's original piles for the Fair Department Store, Chicago (1890) awaiting new pile caps and foundation structure. Location: Dearborn Street, Chicago

Source: Amec Construction Management

The Fair Department Store on Dearborn Street in Chicago, completed in 1892, is known for being one of the first frame buildings constructed entirely in steel, rather than wrought iron, with a non-load-bearing masonry façade supported at each storey. It was designed by the American engineer William LeBaron Jenney (1832–1907) who had been educated at the École Centrale des Arts et Manufactures in Paris and was a pioneer in bringing European building practices to Chicago in the late 1860s. He was one of several innovative engineers working in Chicago who revolutionized the design of foundations over the following few decades. The soft soil beneath Chicago was a particular challenge to engineers wanting to construct buildings of more than 10 storeys. Jenney's use of steel and a lightweight façade allowed traditional spread footings to carry buildings of up to 15 or 16 storeys, where previously 10 had been the limit.

Although the foundations of the Fair Store were designed to carry 16 storeys, the developer decided to limit the building to 9 storeys in its first phase. In 1923 the original spread foundations beneath each column in the single-storey basement were painstakingly dug out by hand and replaced by a 3-storey reinforced-concrete caisson basement, and 2 more storeys were added to the building. The original structure was demolished in 1985 leaving the caisson foundations: a 1m thick spread footing that had supported the original perimeter wall and a 3-storey hole in the ground.

When plans were made in the late 1990s to redevelop the site and construct a new 37-storey tower, it was realized there would be a great saving in both cost and time if the 1923 caisson foundations could be reused. Since the planned column locations did not coincide with the grid of the original building, a 1.5m thick raft of reinforced concrete was constructed in the old basement to transfer loads from the new building to the old foundations.

Date: Original nine-storey building 1890–92
Client: The Fair Department Store
Engineers and architects: William LeBaron Jenney
Date: New 37-storey building 1999–2002
Client: The Beitler Co. and Prime Realty
Architect: Ricardo Bofill and DeStefano and Partners
Construction Manager: Amec Construction Management
Refs: Amec Construction

5 Design Guidance: Building Structure

Reuse, reclamation and recycling in the structure of buildings

The structural elements of a building represent a large proportion of the total mass and for that reason alone their reuse is likely to bring significant environmental benefits. Not only will it reduce the need for new materials, it will help reduce demolition waste that might otherwise go to landfill. Reusing a building in situ brings the added advantages that a good deal of construction activity is avoided, including traffic, and the building will fit harmoniously into the cultural and heritage context.

Reuse in situ

The value of reuse in situ

The principal benefit of reusing the structure of a building is that this will lengthen the life of the building, and generally at a lower cost than demolition and rebuilding. Of equal importance may be preserving the original construction materials that contribute so much to the cultural or heritage value of a building. This may arise for three reasons:

1 preserving the use of locally produced materials in vernacular buildings, such as brickwork or timber-framed construction;
2 the use of construction materials typical of a building type, such as cast or wrought iron in a 19th-century factory or railway station;
3 the use of construction materials typical of a particular era, such as reinforced concrete in many art deco buildings from the 1930s.

Redeveloping an existing building may have particular benefits for a developer or client:

- Old buildings, especially those used as factories, have large floor-to-ceiling storey heights providing plenty of room for raised floors or suspended ceilings and modern air-conditioning systems. Industrial buildings were probably designed to carry floor loads in excess of those imposed in modern office or residential use.
- Many old buildings are suitable for adding a new storey, especially when an industrial building designed for high floor-loadings is redeveloped for a new use with lower loadings, such as a residential building.
- It can be relatively straightforward to remove and replace a non-load-bearing building façade without having to make significant changes to the structural frame. This can be especially valuable if the building is high-rise in an area where, today, planning consent would not be given to such a tall building (Figures 5.1 and 5.2).

Structural appraisal

The first question a potential developer of an existing frame building needs to answer is whether the building will be able to carry the loads that its new use will impose. Generally the answer will be 'yes', but a structural engineer will need to undertake a full structural appraisal to prove this.

All structural/load-bearing components of buildings can be reused as long as they are in sufficiently good condition to perform the duty required of them. Their suitability for reuse will depend on an assessment of their condition – a process generally known as the structural appraisal and is usually applied to existing buildings when they are assessed for a change of use (BRE, 1991; Beckmann, 1995). Every country has its own guidance for structural engineers on how to undertake this process, either in the form of official codes or books on building conservation. The basic philosophy of the approach

Figure 5.1 Winterton House: The naked steel frame and core awaiting new floors and façade

Source: Whitbybird Engineers

Figure 5.2 The new façade is partially supported from above by a new structure fixed to the top of the original reinforced-concrete building core

Source: Bill Addis

is as follows (and is similar to the strategy described in Chapter 4 for assessing the potential for foundations and groundwork to be reused).

The appraisal procedure consists of the following stages:

1 Search for and desk study of documentary evidence about the building.
2 Detailed investigation/survey of the building.
3 Assessment (including structural calculations) of the suitability of the structure for the intended use (Figure 5.3).
4 Recommendations for work on the structure (repair, strengthening etc.).

Investigation/survey of the building

The purpose of a thorough investigation is to establish exactly what is in the building and the arrangement of the various components and hence confirm any drawings that may exist. The activities of such an investigation will usually include the following:

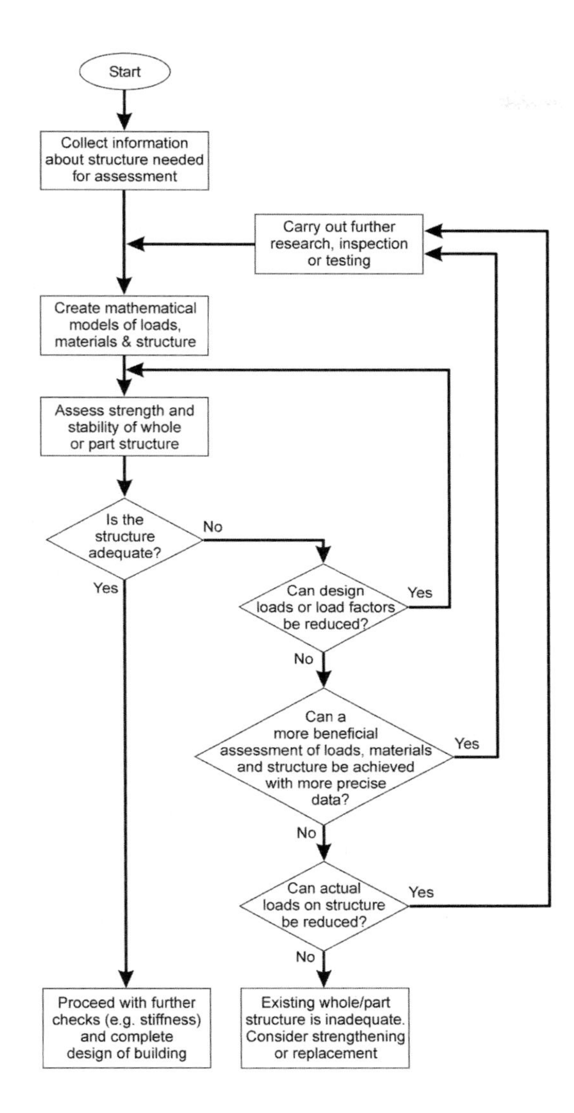

Figure 5.3 Flow chart of the process of appraisal/assessment of a structure

- undertaking a condition survey of the whole building (a visual survey);
- undertaking a measured survey of the whole building;
- undertaking a structural inspection to establish the construction materials, the structural system, and the form and cross-section of individual structural elements;
- establishing the position and details of connections between structural elements, possibly requiring removal of covering material such as concrete, plaster or fireproofing (for example beam/column, and floor beam/wall);
- finding and identifying any structural defects (including corrosion, rot etc.);
- establishing the nature of any alterations that have been made to the original structure;
- establishing the dead loads carried by the structure (e.g. floor screeds, building services);
- undertaking an assessment of structural materials (see below);
- digging pits to expose the foundations;
- undertaking other geotechnical investigations to establish current ground conditions.

Materials assessment

Before assessing the capacity of the whole structure to carry loads, it is essential to determine the nature of the materials of which the load-bearing elements are made, their engineering properties and their condition. This will include the following:

- Identification of the type of material (cast iron, wrought iron, steel, type of timber etc.). For metals it may be necessary to commission a metallurgical investigation to determine its crystalline structure and the constituents of an alloy.
- Identification of the structural material properties – density, yield/ultimate strength in tension and compression, stiffness, brittleness, and so on. This will require the removal of samples of the materials for testing in a materials laboratory.
- Investigation of the condition of the structural materials (e.g. masonry – evidence of frost damage, adequacy of mortar joints), timber (wet or dry rot, infestation, splitting, knot holes), iron/steel (corrosion, drilled holes, condition of welds), concrete (staining indicating internal corrosion of reinforcement or leaching of salts from aggregate, reinforcement revealed by spalling due to corrosion or frost damage), and so on.

The result of the materials assessment will be:

- Materials properties for use in the structural assessment and design calculations.
- Prediction of the likely behaviour of the material for the proposed life of the building (corrosion, creep, decay, behaviour in a fire etc.).
- Identification of any remedial work to be undertaken to improve performance, for example treatment of corrosion, rot or infestation, fire protection, and so on.

Assessment of the structure

The first goal of the assessment is to establish the dead loads that the current structure is carrying. It is also necessary to understand how the structure works – how it carries the dead and imposed loads and the (horizontal) wind loads.

Next, the assessment must consider the alterations that are being proposed in the reuse of the structure, especially:

- different or additional loading on the structure and foundations;
- changes to the ways the structure carries loads to the foundations (the load paths) – for both gravity loads and wind loads (i.e. stability);
- changes in the exposure of the structure, maybe leading to corrosion or rot;
- changes that may affect the fire resistance of the structure.

The outcome of this stage of the assessment will be one of three possibilities:

1. The structure (and proposed alterations) will be fit for the new purpose.
2. The structure is unfit and effectively beyond salvation (i.e. only at great expense).
3. The structure is apparently unfit for the new purpose, but a combination of more sophisticated structural analysis and minor changes to the design/construction can make it fit for purpose.

Dealing with the last of these outcomes often requires great ingenuity and good engineering judgement based on experience. Good guidance on these processes of structural appraisal is available (for example BRE, 1991; CIRIA, 1994; IStructE, 1996; Bussell, 1997).

Warranty

The reuse of all structural elements of a building will require the involvement of a structural engineer. After properly following the appraisal procedure outlined above the structural engineer will be able to provide the necessary surety that the structure will be sound, just as he or she would for a newly designed building or structural element.

As with the design of a new building, the proposed reuse of a building for a new purpose will need to be approved by a local regulatory authority. This will apply not only to architectural issues but also to the design of the structural elements. Apart from providing assurance that the proposed reuse complies with local regulations, the local regulatory authority can be a valuable source of guidance about other buildings in the area, that may be of similar construction, and local geotechnical features. Such authorities often tend to have their own views on their local built environment, and practices considered acceptable in one city or region may be considered unacceptable in another.

Reuse of a building in a new location

There are occasions when it may be considered appropriate to move an entire building to a new location. Usually this will be when the building is of great historical value and needs to be moved to make way for a new road, railway or other major development. Two solutions are possible – dismantling the building and re-erecting it, or moving the building intact. While both are occasionally undertaken, it should be observed that such moves are not done with the intention of minimizing environmental impact or conserving materials.

Disassembly and re-erection

Many types of temporary building are designed to be dismantled and re-erected in a new location – circus tents, temporary exhibition buildings and moveable stages for music or theatre performances. Such buildings are designed with regular disassembly in mind, and do not fall within the scope of this book.

Old timber-framed buildings are particularly amenable to dismantling and reassembly because traditional timber jointing methods are inherently reversible. In such cases the timbers need to be carefully labelled to ensure their re-erection in the same way because each timber elements was made to fit a unique position; likewise with brick walls built using lime mortar. They can be dismantled brick by brick, and rebuilt after cleaning the bricks. Windows and tiled roofs are generally easier than brickwork to remove and rebuild.

Moving the building

If a building is constructed using materials that are not possible to dismantle and rebuild, it can be viable to move the building intact. Although normally done for old and fragile buildings, it has sometimes been done for buildings constructed in reinforced concrete. One such example was the Museum Hotel in Wellington, New Zealand.

Case Study: Museum Hotel, Wellington, New Zealand

In 1993 the New Zealand government decided to locate its new national museum on a site adjacent to the harbour, part of which was occupied by a five-storey hotel constructed in 1986. When it was discovered that the government planned to demolish the building rather than use it, a number of businesses showed interest and persuaded the government to allow tenders for moving the building to a new location outside the proposed museum site. The

Figure 5.4 Rail tracks installed at ground level

Source: Dunning Thornton

winning client engaged engineers Dunning Thornton as the lead consultants.

The entire structure weighed around 3000 tonnes with a maximum column loading of 160 tonnes. To get to the new location the building had to be moved around 80 metres south east and then 70 metres south west. This was achieved by transferring the load in each column and the two full-height concrete shear walls onto a steel frame supported by a total of 90 railway bogeys. Meanwhile foundations were built in the new location and two sets of rail tracks laid (Thornton, 1994) (Figures 5.4 and 5.5).

After four months of preparation the first leg of the move was completed in a day. Hydraulic rams provided the 15 tonne force to move the building at 20 metres per hour during the push stroke. While the bogeys were transferred to the second set of tracks at right angles, the building settled around 40 millimetres and a force of 70 tonnes was needed to push the building out of this depression (Figure 5.6). After the second move was completed over a weekend, the loads in the columns were transferred from the bogeys to the new foundations and the new Museum Hotel opened for business just five months after the work had begun.

Reuse of salvaged or reconditioned products and reclaimed materials

Opportunities for using salvaged components and reclaimed materials

The value of reusing structural components from another building can be environmental, such as

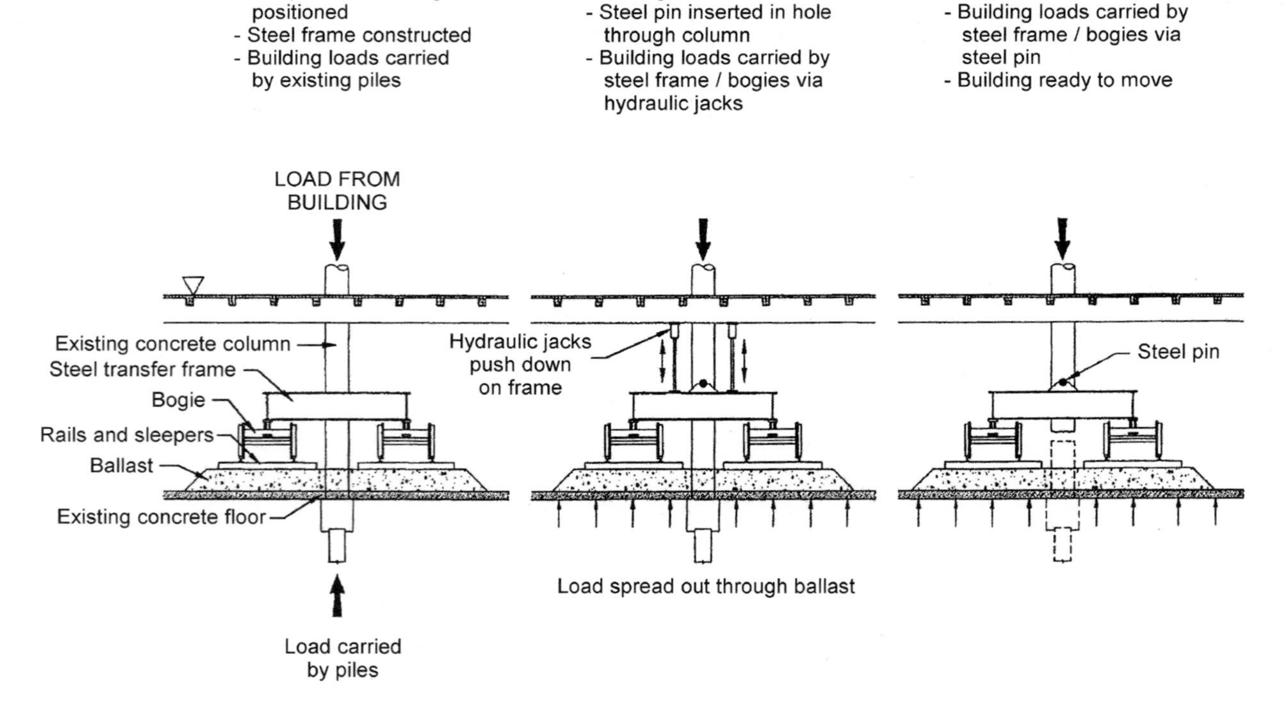

Figure 5.5 Method for transferring the weight of the Museum Hotel from the foundations to the railway bogies

Source: Dunning Thornton

Figure 5.6 The Museum Hotel on the move

Source: Dunning Thornton

using less primary resources and reducing impacts from manufacturing processes and transport. It may also be commercial – the cost of second-hand (reclaimed) structural timber can be less than the new equivalent. In the future it is likely that commercial pressures will encourage more structural steel to be reclaimed as the cost of new steel rises.

The reasons for reusing individual structural elements will not be quite the same as for reusing a whole building, and vary with the material and type of construction:

- The structural element may be visible and need to match other materials in a building (e.g. an ancient timber beam, 18th-century brickwork, a cast-iron column in a mill).
- Only very few structural members are needed (maybe only one) and the precise size is flexible (e.g. a steel girder to bridge an opening made in a masonry wall, a few steel girders or timber beams to bridge across a hole cut in a floor during a previous refurbishment of a building).
- An architect's desire to use reclaimed structural components in the name of 'sustainable construction'.

Masonry

The main reason why masonry materials and components would be reused from a different location is to match the materials used in an old building when being repaired or restored or when an extension is being built. Many planning authorities are strict about the use of construction materials in conservation areas or for the repair of buildings important to the local or national heritage. Great care is needed to match the colour, size and age of bricks and stone.

Timber

Traditional timber-frame buildings form an important part of the heritage of most countries and many such buildings are restored and reused and, for this reason, great care is taken in preserving them. Unfortunately, it sometimes happens that an old timber-framed building finds itself in the 'wrong place' when a modern town centre development is proposed or a new motorway or railway is constructed. In such cases, if the old building is of sufficient cultural importance, it is possible to move the entire building. A timber-framed building can relatively easily be dismantled, all the components, having been first clearly labelled and re-erected on a new site. The size and robustness of the major components used in timber construction is such that they are easy to reclaim when a building is demolished and can easily be used in the repair of buildings of a similar age.

Cast iron

It is rather unlikely that a building designer would choose to use reclaimed cast-iron structural elements salvaged from a demolished building since the material is nowadays not considered a suitable structural material (mainly because it is brittle). An exception to this general rule might be when a building is being partly demolished and some beams and columns could be salvaged to replace damaged ones in the part of the building being retained. This can be of great value to recreate the visual impression of the original construction.

Wrought iron

Wrought-iron beams and columns were not only made by riveting together several small sections, they were also usually riveted together at their connections. The difficulty of dismantling such components economically and without damage would usually outweigh the value to be gained by their reuse. Wrought-iron beams and columns were seldom left exposed and there would be no value in reusing such components only to cover them with fire protection.

Wrought-iron roof trusses, however, were often left visible in the upper storey of buildings and can be dismantled, refurbished and reused. It is unlikely (though possible) that an entire roof would be dismantled and reused elsewhere. However, in the case of partial demolition of a building, some roof trusses removed from a redundant part of the original building could be used to replace damaged ones in the retained part in order to maintain the visual impression of the original building. In the absence of a heritage reason for reusing wrought-iron roof trusses, there is little or no point in reusing old wrought-iron components.

Steel

There are strong environmental reasons for reusing steel sections: by reusing a steel beam there is a twin saving of energy – the energy needed to melt the steel in the recycling process and the energy expended in making the new steel beam that would otherwise be needed.

In principle all rolled-steel sections and elements of roof structures can be reused, as long as they are judged to be structurally adequate for their proposed purpose. The value of doing so will be determined by the availability of suitable sections in adequate quantities, and the convenience (or otherwise) of adapting the components found to the storey height and column spacing of the new building.

Reinforced concrete

At present it is likely that only precast concrete building elements might be salvaged and reused. In fact this would merely be the reverse of the semi-industrialized approach to building construction that first led to the development of precast concrete systems in the late 19th century, and which has undergone several revivals in the intervening century. In recent years in The Netherlands, for example, a number of industrialized building systems have been developed with the intention that they should be able to be dismantled and re-erected elsewhere. In principle, there is no reason why an imaginative design team could not plan to dismantle and reuse the pieces of an existing building with precast concrete elements.

Availability (quantity, size and shape)

Virtually all structural components are made precisely to suit their purpose in their original location – both their load capacity and stiffness (namely cross-sectional shape), and how they are fitted into the building (namely their length, depth and method of fixing to other building elements).

Nevertheless, many structural members are of similar size and shape and potentially reusable after removal from a building. It will probably be necessary to refurbish such components, for example by removing paint, fire protection or other building materials adhering to the surface. Timber and steel beams and columns can easily be cut down to the desired length and new end-connections can be machined. Repairs can be made (such as filling holes) and components can even be lengthened by joining new sections on.

As with all reclaimed components and materials, it will be essential to find a source of suitable numbers and sizes *before design begins*. For example, if a sufficient number of steel beams of roughly suitable depth and length is found, the spacing of columns will need to be chosen to suit.

Structural appraisal

The reuse of structural elements from a different location is subject to all the same issues as reuse of structural elements in situ. A structural appraisal procedure will need to be followed to prove the items are capable of performing the duty required, and will be sufficiently durable to meet the building's needs.

Warranty

As with reuse in situ, the performance of reclaimed structural components will be assured by the structural engineer who undertakes the appraisal.

Recycled-content building products

The differentiation between the reuse of a structural component and the recycling of a structural material is perhaps open to some debate, for example whether a brick is a structural component or a structural material. For the sake of consistency in this book 'recycling' is confined to the manufacture of a new material using basic ingredients that have already been used once. Hence, using a salvaged brick is an example of 'reuse' not recycling. If a brick were manufactured from a clay made from pulverized bricks and then baked in a brick oven, then it would be a brick made from recycled materials. Similarly, using a piece of reclaimed timber to make a floor beam is called 'reuse', while chipboard or blockboard made from post-industrial or post-consumer waste timber is considered to be a product with recycled content.

Another possible confusion with recycled-content structural materials arises with steel and other metals. Almost all 'new' steel and aluminium in fact contains recycled material, and virtually no metal is discarded as waste and sent to landfill. There is thus no value in a designer specifying a steel or aluminium component with recycled content since doing so would not prevent any material going to landfill.

Generally speaking, then, the only structural components likely to be made from recycled materials that would otherwise be waste and be sent to landfill sites are those made from concrete (e.g. blocks) with recycled content.

Masonry (load-bearing and non-load-bearing)

Masonry construction generally consists of a cuboidal block, made of brick, natural stone, concrete or even glass, stacked to form a wall or a column, with mortar-filled joints between the blocks. The mortar serves both to provide a filler material that spreads loads evenly across the full area of the block and also to bond adjacent blocks to one another. Most masonry construction consists of one of the following range of materials together with a suitable mortar:

- stone walls made from dressed stone ('ashlar' or 'dimension stone') or undressed stone (rubble walling, drystone walling);
- other major elements of stone buildings, including arches, vaults and buttresses typically found in churches and cathedrals;
- staircases made of stone (so-called 'cantilever' stone staircases are discussed in Chapter 6);
- brickwork using fired clay bricks;
- blockwork made using blocks of ordinary or lightweight concrete;
- exterior cavity walls with one skin of brickwork (external) and one of blockwork (internal);

- shaped, precast concrete blocks or components (for lintels etc.);
- reconstituted stone – concrete blocks or mouldings made using high-quality aggregate, often polished to resemble ashlar;
- partitions made from glass blocks.

Apart from the main types of block used in large numbers, there is a variety of bespoke components incorporated into masonry walls for specific purposes, for example:

- lintels, sills, mullions and transoms, and jambs;
- copings and cornices;
- pier and chimney caps;
- finials, corbels and small panels;
- stone staircases including treads, landings, handrails and cappings;
- balustrades;
- corbels and padstones.

In addition, a wide variety of products are used in association with the masonry components, including:

- bar reinforcement for grouted cavity, quetta bond and pocket reinforced brick/blockwork;
- wall ties, anchors, starters and connectors;
- damp-proof courses (DPCs) made of slate (in traditional buildings) or flexible materials (polymers, bitumen-impregnated textile etc.);
- cavity trays;
- rigid sheet and various injected cavity-insulation materials;
- air bricks and cavity-wall ventilators;
- flue linings;
- various 'fix only' items that are built into the walling, but are not an integral part of it, including built-in windows and doors, ends of joists and joist hangers, holding down straps, ends of floor and roof beams, fireplace surrounds, and so on.

Many types of building have been and are constructed from masonry: many Roman buildings, medieval castles and cathedrals, magnificent country and town houses from the 16th to the 19th centuries, workers' cottages in 19th-century industrial towns, as well as millions of modern houses. This range and number of buildings is testament to the durability of masonry and to the ease with which it can be adapted to new fashions and needs, as well as the ease with which it can be maintained in good condition and repaired.

The principal difference between load-bearing and non-load-bearing masonry is that the former is made of stronger ingredients. Generally speaking, any masonry used in a building that has a structural frame of timber, steel or reinforced concrete, will be non-load-bearing. Before around 1900, many buildings were constructed with outer, load-bearing masonry walls and internal columns of cast iron or, up to the 1820s, of timber.

Mortars for masonry construction generally fall into two categories – traditional lime mortars and modern cement mortars. Lime mortars harden very slowly (years), which allows the masonry construction to adapt to small movements in a building due to settlement of foundations or shrinkage of structural timber without causing cracks. Lime mortar also bonds very lightly to the brick or stone. This has the advantage (for reuse) that individual stones or bricks can be separated with relative ease and without causing damage. Modern cement mortar (common since around 1900) hardens very quickly (days) and becomes brittle when hard. Movements in the fabric of a building will usually lead to cracks in both the bricks and mortar of the masonry. The bond between the cement mortar and the bricks or blocks is very strong – sometimes stronger than the bricks or blocks themselves – and so it is relatively difficult to separate individual bricks or blocks. Also, cracks in masonry with a high-strength mortar, caused by foundation movement, for example, are likely to run through the bricks or blocks rather than be confined to the mortar.

Reuse in situ

Structural appraisal

The general procedures described at the start of this chapter should be followed for load-bearing masonry. Assessing the condition of structural masonry is of paramount importance in dealing with nearly all buildings constructed before the late

19th century. Much excellent guidance is published within the heritage industries of many countries and this must be consulted before assessment is undertaken. The most convenient way of ensuring the correct guidance is consulted is to employ structural engineers with a proven and appropriate track record of working with existing masonry structures (Figures 5.7 and 5.8).

Materials assessment

The load-bearing strength of masonry can often be judged by visual inspection. Stones or bricks that have been damaged by impact can be identified and a judgement made as to whether the integrity or the load-bearing cross section of the unit have been affected. Both stone and brick can be eroded by the action of wind and rain that can significantly reduce the cross-section in a very old building. They can also be damaged by frost when water that has penetrated into surface micro-cracks expands on freezing and causes surface layers to spall.

Structural assessment

Most masonry is not highly loaded. Even at the base of a cathedral spire the stone is loaded to less than a tenth of its compressive strength. Nevertheless, a structural engineer may judge it necessary to measure the strength of a section of masonry or the strength of the stone used. This will enable the engineer to increase the level of confidence with which the stability of the structure can be predicted. Samples for testing will need to be removed from a location that is not critical from a load-bearing or visual point of view. A chemical analysis of a stone, together with a geologist's guidance, can help identify its precise origin and even the actual quarry where it was hewn. This will enable stones to be cut for use in repair or new construction that must match original masonry.

Remedial work

The appearance of masonry is one of its particular charms and a wealth of guidance is available within the heritage industry of most countries on maintaining, cleaning and repairing masonry.

Whether of stone or brick, masonry construction uses naturally occurring materials that weather over time, and usually change colour according to the type of pollution in the atmosphere that contaminates the rain falling on a building. Discolouration can be reversed by cleaning, though great care is needed to ensure that the process, especially if chemicals are used, does not itself cause discolouration or damage the stone or brick. It is always wise to clean a test area away from the most visible parts of a building façade.

The advantage of all masonry is that damaged stones and bricks, whether a large area or individual units, can be removed and replaced. As long as care is taken to match material, colour, dimensions and the thickness and type of mortar, repairs can be nearly invisible, especially after a few years of weathering. It will be necessary to undertake laboratory tests to establish the chemical composition of the mortar to ensure that an appropriate matching mortar is used for the repairs. It will be necessary to provide temporary support if removal of stones or bricks affects the stability of the structure or would cause increased deformations.

The structural performance of reused masonry construction will be assured by the structural engineer who undertakes the appraisal and any alterations. Contractors will be responsible for the quality of repair, remediation work or cleaning.

Reuse of salvaged or reconditioned products and reclaimed materials

The large number of masonry buildings constructed means that large numbers of such buildings are being demolished all the time. The opportunity to reuse masonry depends very much on the ease with which the individual bricks or stones can be separated and cleaned. Demolition contractors are nowadays well aware of the most valuable items in masonry construction and such materials and special items now quickly find their way into the architectural salvage yards located throughout many countries. The challenge will always be to find masonry of the required type (age, material, colour, size, quantity etc.) to match the materials of the original construction.

So desirable are some items such as stone fireplaces, decorative stone gateposts and stone staircases, that they are often robbed from unoccupied old buildings and sold on to the less reputable architectural salvage companies.

Figure 5.7 Load-bearing stone walls at the East Mill, Huddersfield before refurbishment

Source: Bill Addis

Figure 5.8 The building after renovation as the maths/computer department of the University of Huddersfield

Source: Bill Addis

Reclaimed masonry will usually be sold 'as seen' without a warranty guaranteeing its performance. Fortunately, a visual inspection by an experienced user of masonry will usually be satisfactory to ascertain the quality of reclaimed masonry and its suitability even for load-bearing purposes. If there is doubt about the compressive strength of stones or bricks, individual units can be tested but the results must be used with care – the strength of a wall, vault or column will depend on the strengths of both the mortar and the individual stones or bricks.

Most critical for reclaimed bricks are their colour and size. The colour will vary with the type of clay used and the temperature at which the brick was fired. While bricks are usually approximately the same size, especially in a particular region, variations of just a millimetre or so in height or length can become apparent when new brick work is constructed alongside existing brickwork since the thickness of the mortar joints will need to vary to ensure the bricks themselves line up (Figure 5.9).

Masonry has a high transport cost and reclaimed material tends to be stored within 50 or 60 miles of the demolition site from which it is salvaged. This has a particular benefit for bricks since they are likely to be made from clay from the same source

Figure 5.9 Sheltered accommodation for the elderly built from reclaimed bricks, Tewkesbury, UK

Source: Salvo

and hence match other brickwork in colour and texture; likewise, salvaged stone is likely to be from local quarries.

Recycled-content building products

Bricks can be made with recycled content using a variety of post-industrial waste materials, including colliery spoil, dredged silt, pulverized fuel ash, blast-furnace slag and sewage sludge. The percentage recycled content mixed with virgin clay varies considerably according to the materials being used and the type of brick, but can exceed 90 per cent. Further information on recycled-content bricks is available from national trade associations such as the British Ceramic Confederation or the Brick Development Association in Britain.

Concrete blocks can be made using a proportion of reclaimed or recycled materials including PFA and blast-furnace slag. Further information is available from national trade associations such as the Concrete Block Association in Britain.

A great many types of block, both in regular sizes and in bespoke shapes used for mouldings and decorative architectural features can be made from concrete including recycled aggregate. Such products can be specified and purchased as new products.

As with all RCBPs, it may be important to know the recycled content and to be sufficiently confident in the quoted proportion if a certain recycled content is a requirement of a building brief.

Structural frame: Timber

Timber is an extremely versatile construction material. Its uses are difficult to categorize precisely, but for practical purposes can be considered under two main headings:

1. load-bearing uses that today require the design or appraisal skills of the structural engineer;
2. semi-structural elements such as timber studding for partitions and prefabricated wall panels.

To some extent the distinction between these categories is one of scale. In the first category we have:

- large-section, load-bearing timber elements forming the columns, beams and roof trusses of many traditional buildings such medieval halls and Tudor half-timbered houses;
- fabricated load-bearing elements such as glued laminated timber beams and portal frames (glulam);
- various structural components of roof trusses (rafters, purlins, posts, struts etc.);
- prefabricated timber frames including trussed rafters;
- floor beams.

The second category comprises elements made from smaller sections of timber and timber products such as plywood, block board and other panel products. This includes:

- timber studs, battens, plates, bearers, grounds, noggings, packings, firrings and fillets;
- prefabricated wall panels with ply sheathing;
- beams and frames made using plywood or hardboard webs and gussets;
- timber valley boards, parapet gutter boards, bargeboards, fascia boards and eaves soffit boards to pitched or flat roofs.

In addition to the main timber elements are many accessories, usually made of metal, including:

- bolts, metal connectors, straps, nail plates, wire or metal rod bracing;
- hangers, shoes, wall-plates and bearers.

Timber-frame buildings have been constructed in a variety of ways since medieval times and many such buildings date from before around 1900 are protected as an important part of the heritage of many countries. For this reason numerous techniques have been developed for preserving and restoring these buildings. Although timber is prone to decay from wet and dry rot, a well-built, well-maintained (and well-ventilated) timber frame has an almost indefinite life. If partial decay or damage is discovered, many means of treatment and repair are now available.

The larger timber sections used for load-bearing purposes in timber-frame buildings, floors and roof trusses are all relatively easy to dismantle without damage, and their size also means they can be cut, planed or sanded to remove surface damage or discolouration and still leave a piece of timber large enough to use again.

Timber floors, comprising timber joists and floor boards, are found in large quantities in nearly all houses built before the 1950s as well as many large 19th-century buildings and in nearly all buildings constructed before 1800. Timber beams are subject to creep under long, continuous loading and old floors may sag noticeably.

Reuse in situ

The timber elements in the structure of an existing building must be appraised using the standard approach to structural appraisal summarized above. The appraisal of a timber structure consists of two main stages:

1. Survey – review of documentary information and a visual inspection to obtain information on structural form, condition, loading and environmental factors.
2. Appraisal – assessment of the potential for reuse, including strength, stability, serviceability, durability, appearance and accidental damage.

Structural appraisal

Details of the methodology for appraisal of any existing structure are given at the start of this chapt er. More detailed guidance is also available for timber structures in Ross (2002) and Yeomans (2003).

Material assessment

The possibility of reusing timber in situ depends on its condition and fitness for purpose. All timber can be damaged by wet and dry rot and by infestation of various organisms. A variety of tests need to be carried out to establish the condition of the timber and its suitability for reuse in situ. Various investigative techniques are available for assessing the physical and chemical properties of the timber and hence the potential for reuse; these include the borescope, moisture content meter, auger drilling, the Silbert drill, various non-destructive testing methods, dendrochronology and radiocarbon dating.

Timber used in construction is often treated to prevent rot or woodworm infestation. Some treatments involve the use of hazardous substances (e.g. arsenic), which may make reuse or recycling more expensive or impossible. Timber research institutes, such as the Timber Research and Development Association (TRADA) in Britain, can advise on laboratory testing for timber treatments if necessary.

The main consideration when reusing timber is the strength grade. The strength grade will primarily determine whether members are suitable for reuse in load-bearing applications. The strength grade of timber (both virgin and used) is typically affected by natural strength defects such as knots, slope of grain, rate of growth, wane, fissures, resin pockets, bark pockets and distortion.

Other issues to be established when assessing timber for reuse include:

- type of timber (it is important to know whether it is softwood or hardwood, but if possible the species should be identified as different species have characteristic properties that are useful in determining the potential for reuse);
- the age of the timber;
- moisture content;
- dimensions and surface finish;
- knot and peg or bolt-holes;
- defects associated with used members (fungal and/or insect attack, instability, deterioration, structural inadequacy, deformation, noise transmission and vibration of floors, and accidental damage such as overload, impact, explosion, fire and strength-reducing alterations);
- presence of other materials that may impact the aesthetics or long-term performance of members or may have health risks (such as asbestos, bricks, cardboard, concrete, dirt, rubble, stone, drywall, fibreglass, glass, linoleum, nails, screws, metals, plaster, plastics, creosote, heavy metals (e.g. chromium, arsenic), paints, pentachlorophenol, resins, varnishes, stains, phenol formaldehyde, preservatives, diseases and fungicides);
- variable properties and performance (depending on type, quality and previous usage);
- durability;
- fireproofing;
- applied surface treatments;
- storage conditions (deterioration may occur in storage).

The outcome of the material assessment will be the structural data needed to undertake the structural assessment, as well as recommendations for any remedial work or treatment needed to ensure the long-term life of the material.

Structural assessment

Investigative techniques may be required during the appraisal process in order to determine certain physical and chemical properties of the timber and hence the potential for reuse. Structural timber members should first be inspected visually for the following:

- the condition of connections and supports in a masonry wall, especially signs of decay or rot;
- longitudinal splits in the wood (shakes) (which actually may not be too detrimental);

- evidence of previous repair.

To assess the strength of timber members and the entire frame that they make up, it will be necessary to ascertain the following:

- dimensions (or 'scantlings') of members (which may be irregular), including knot holes;
- types of joints and connections, including wooden pegs, wedges or bolts;
- dimensions and arrangement of any iron tie rods, hangers, shoes and bolts (especially in roof trusses).

To assess the stability of the entire structure it will also be necessary to ascertain the strength and stiffness of structural connections as well as the contribution of any composite action between the timber frame and the walls (either load-bearing masonry or brick or mud infill) and the floor.

To ascertain these properties, it will probably be necessary to carefully remove ceilings, floorboards or portions of walls. Such damage can usually be repaired satisfactorily.

Even if a conventional analysis of structural behaviour would seem to indicate that such beams or columns do not satisfy modern design codes, a more careful, bespoke analysis may enable a higher performance to be justified, especially if design loads can be reduced from the normal (high) modern levels.

Remedial work

Both structural and non-structural timber in an old building can be reconditioned by simple, but often time-consuming processes, including:

- removal of nails and screws;
- removal of decayed or damaged parts;
- removal of paints, varnishes or plaster;
- cutting or planing to desired size, as with new wood;
- filling knot holes and redundant peg or bolt-holes;
- member replacement (in the case of entire frames);
- part replacement (generally ends);
- sanding or other finishing;
- treatment to prevent decay;
- normal fireproofing to minimize flame spread.

Nails are a major problem in the recovery and reuse of old timber because they can be hard to remove and effectively contaminate the wood. Clean-headed nails can be easily removed, but others may require chiselling out. Larger beams used to support masonry openings or joists tend to have fewer nails, and are therefore more valuable per cubic metre. Where sections are smaller, nails become more frequent, and the timber less valuable. Timbers containing large quantities of nails may be uneconomical to reuse due to the time required to remove them; however, many drum grinders can cope with nails.

Although screws in good condition are easy to remove, their heads are often damaged or clogged with paint. Removal of screws with damaged heads is much more difficult than removing nails and the process will usually result in greater damage to the timber.

Old timbers may have been affected by woodworm or other infestation. Where the infestation is live, the timber will require treatment before use. If the infestation has died out, then the timber can be used safely although it may not have the strength needed for structural use.

If the structural appraisal concludes that a timber frame is unable to be reused in its current state, there are many types of remedial work that can be undertaken. These vary according to the nature of the deficiency, for example timber damaged by decay or infestation can be cut out and replaced. A timber beam judged too weak to carry the loads imposed on it can be strengthened and stiffened by fixing reinforcement to the outside of the structural element. As well as steel, glass fibre or carbon fibre can be used today.

While old floorboards are often stained or worn on the surface, they are usually much thicker than modern boards (typically 25mm) and they can be planed or sanded to restore them almost to their new condition. While the failure to remove all nails is not likely to affect the performance of floors, nails will damage planing equipment and saws.

A timber floor can be stiffened or strengthened by replacement of individual beams or even, for particularly fine floors, by fixing steel, glass-fibre or carbon-fibre reinforcement to the underside of joists. This method is particularly useful for strengthening or stiffening timber floor beams when the headroom in the room beneath is particularly low.

Warranty

As with new construction, the structural engineer responsible for the appraisal and any remedial work will be able to warranty the structural performance.

Reuse of salvaged or reconditioned products and reclaimed materials

Availability (quantity, size and shape)

There are many good opportunities to reuse structural timber. It is a common building material and members with superior quality, particularly those recovered from old buildings, are highly valued for reuse. Further to this, the recovery infrastructure is growing and markets for reclaimed timber are improving.

It is in the nature of timber-framed buildings that they can be entirely dismantled and re-erected at a new location. This might be out of necessity such as when an old building stands in the way of a new development (e.g. a new motorway or railway); or an old building might be dismantled and re-erected in a museum to preserve it for posterity.

Most timber beams, columns and components of roof trusses no longer needed for their original purpose can be dismantled and reused in other buildings. Generally such components would be used to repair a building of similar vintage rather than to construct a new building.

There are many reasons why salvaged old timber is highly sought after today:

- it is often available in sizes, lengths and species that currently can be difficult to obtain;
- it is dense, tight-grained wood;
- it is often free from knots and defects;
- it is well-seasoned and, hence, dimensionally more stable.

Timber for non-structural purposes is likely to be more readily available than structural timber, due to the more stringent quality and performance requirements for load-bearing applications.

In addition, while beams and joists are frequently recovered for reuse, smaller dimensional timber is not always recovered for reuse. This is because it can be difficult to salvage from existing assemblies without substantial damage. If timber is to be salvaged from a building due for demolition, it will be important that the design team provides guidance to the demolition contractor prior to design of the new building.

The salvaging of timber joists and floorboards from old buildings, when they are fully or partially demolished, and their reuse in the refurbishment of other buildings is now common and markets in such components thrive in many countries.

There are numerous firms that specialize in the salvage and resale of structural timber. Demolition firms often have recycling arms, which will be able to supply timber for reuse. Alternatively, local recycling merchants may be found by contacting large demolition firms. Online searches using keywords such as 'salvage', 'timber', 'structural' and 'reuse' will provide many suppliers of reclaimed timber. For example:

- The Waste Book (www.recycle.mcmail.com), which is a guide to recycling and sustainable waste management for businesses and organizations in London and south-east England, contains a list of companies and community organizations that can make productive use of wood of various types, giving it a second life.
- In the US, the Washington State Department of Ecology's Sustainable Building Toolbox web page (www.ecy.wa.gov/programs/swfa/cdl/index.html), provides direct links to online material exchange services, retail outlets and re-milled timber suppliers.

It may also be possible to purchase timber from demolition sites or salvage yards. In the UK, suppliers may be found on the following websites:

- Salvo directory of live and forthcoming demolition projects (www.salvo.co.uk);
- BRE/DETR materials information exchange (www.bre.co.uk/waste).

Timber is a commodity within a competitive market. Compared to virgin material, much more testing and refurbishment must be done to reclaimed timber in order to determine whether it is suitable for reuse in load-bearing applications. This can result in costs above market prices. Generally, the cost of good quality reclaimed timber bought from salvage yards is likely to range from 60–90 per cent of the cost of new timber. If the selected reclaimed timber is more expensive than new material, the designer and client must decide whether the advantages (outlined earlier) outweigh the additional costs.

Structural appraisal

Generally, structural timber sections are appraised using the same procedure as for timber in situ. The stiffness and strength can be quite accurately assessed knowing the species of timber and the moisture content. If it is judged necessary, it is also practical to test the quality and properties of individual components removed from a structure, for example using stress grading in which the deflection of a member under load is measured and, hence, its stiffness is calculated. Knowing the type of timber, moisture content and stiffness, a more reliable estimate of strength can be made than relying only on visual inspection.

Remedial work

The remediation of timber salvaged from a building is little different from remediation in situ. The question arises, however, of who does it. It could be part of the demolition contract that timber is de-nailed and cleaned of loose material adhering to it. Alternatively, a reclamation contractor or the salvage yard may undertake this work. Or, if the project team for the new building being designed has identified the timber prior to demolition, it could be appropriate for the contractor to undertake the de-nailing and cleaning prior to removal to a storage site. This would have the advantage that the project team would be fully aware of the sizes, condition and properties of the salvaged timber early in the design process.

Warranty

The structural engineer responsible for the appraisal and any remedial work will be able to warranty the structural performance. Reclaimed timber for non-structural purposes will generally be sold 'as seen' and no warranty will be provided. Hence all timber, whether load-bearing or non-load-bearing, will need to be inspected prior to purchase from a salvage yard.

Recycled-content building products

There are no significant opportunities for using recycled-content products for the load-bearing structural elements of a building.

Structural frame: Iron and steel

The iron and steel structural frame includes a wide range of building construction from the late 18th century to the present day. The main types of construction are these:

Structural cast iron

From the 1790s to the 1860s many thousands of factory buildings and warehouses were constructed with load-bearing masonry exterior walls and internal columns and beams made of cast iron. Some roof structures were also made of cast iron. The floors usually consisted of either a brick jack arch or small cast-iron beams supporting stone slabs or flags. These were the first 'fireproof' buildings and virtually eliminated the use of timber.

Many such buildings have been restored to their former glory, either as heritage building (museums etc.) or as fully functioning modern buildings such as flats, offices and workshops.

Structural wrought iron

Wrought iron was first used from the 1780s in France (1820s in other countries) as a replacement for timber in roof trusses. From the 1850s wrought-iron plates and simple L-sections, riveted together to form larger structural I- and H-sections, began to replace cast iron for beams and, later, for columns. Until the 1880s the external walls of such buildings were generally still of load-bearing masonry. The first examples of fully framed buildings with non-load-bearing walls were built in Britain in the 1850s and in France in the 1860s.

Structural steel

Steel replaced wrought iron from the 1880s and the modern structural frame, with non-load-bearing building envelope, developed rapidly in the US and soon spread throughout the world. Rolled steel I-beams and H-section columns were first used on a large scale in the 1890s.

A number of books give details of these types of construction and these should be consulted before any plans are made to reuse entire buildings or structural components reclaimed from such buildings (e.g. CIRIA, 1994; Bussell, 1997; Friedman, 1995; Swailes and Marsh, 1998).

Reuse in situ

Structural appraisal – cast iron

The structural appraisal of an iron- or steel-framed building must follow the procedures outlined at the start of this chapter.

A structural engineer should be chosen who is familiar with buildings of the type proposed for reuse, both in general and, preferably, in the same area of the country, since much construction has its regional idiosyncrasies. A detailed knowledge of the construction material (cast iron, wrought iron or steel) is also essential (Table 5.1).

The age of a structure and its location will generally give a good indication of the codes or standards used when designing the structure. National codes began to gain currency from around the turn of the 20th century. Before that time, authors of books on structural design and individual firms developed their own good practice guides and manufacturers of iron and steel components issued data about their strength and other properties. Further details of the standard texts available in Britain are given in CIRIA (1994) and Bussell (1997). Some similar information for the US is given in Friedman (1995).

When assessing structural elements made of cast iron, wrought iron or steel for their potential reuse, it is important to understand the different structural properties of the three materials, as shown in Table 5.1.

Table 5.1 Qualitative comparison of key properties of cast and wrought iron and steel

Structural property	Cast iron	Wrought iron	Steel
Tensile strength	Poor	Good	Good – excellent
Compressive strength	Very good	Good	Good – excellent
Ductility, toughness	Poor – brittle fractures in tension	Good	Good

The approach to assessing the suitability of a structure for reuse is similar for all materials. The following illustrative example is for a 19th-century building with cast-iron columns and beams and a masonry jack-arch floor. Further details of this case study are given in Chapter 2.

Materials and structural assessment

Structural components in cast iron are made by pouring the liquid metal into a mould made of sand. Once removed from the mould, the casting can be inspected for defects such as blow holes or inclusions of iron slag. A poor casting can be rejected immediately. The quality of the casting beneath the surface is another matter. It cannot be inspected and the quality of beams and columns are proven by testing each one under load – or rather under a load three or four times as large as it would be expected to carry in the building.

An engineer encountering such components in a building does not have the benefit of having seen the load tests so needs to proceed with caution. It is clearly not possible to proof test all components in situ, for fear of a dramatic collapse. An alternative approach is needed.

The condition survey and structural inspection of the building will establish any unusual problems such as corrosion, accidental damage, damage by fire or previous unsafe intervention in the building (such as removal of columns or load-bearing walls). There can then begin the assessment of the structure and the materials of which it is made.

Initially the engineer's eyes will tell him or her that the beams and columns are able to carry the dead loads (the building is standing). Also, the building may be known to have carried certain floor loads until recently – industrial machinery or goods in a warehouse, for example. This should indicate that the structure is *probably* strong enough to serve its purpose in the proposed new use. Nevertheless, further analysis is needed to increase the confidence of such predictions.

The strength of the cast iron can be assumed to be typical of iron made at the time the building was new and in the same region. Such a figure is likely to be conservative. If this assumption leads to a calculation suggesting that the beam or column is not strong enough, then further investigation is needed (since it probably *is* strong enough). The engineer needs to find out more precisely what is actually going on in the structure of the particular building (rather than using average data).

There may be a number of reasons that could lead the engineer to revise the assumptions upon which the calculations were made:

- Tests on samples of the cast iron in the building may indicate it is stronger than the average value.
- The dimensions of the beams and columns may be different from first estimates – especially their actual lengths that can only be determined by removing some material to inspect connections.
- It is usual to assume that beam/column connections are pin-jointed with no capacity to carry bending moments; on investigation it may be found that the beam/column connections do in fact have some capacity to carry bending moments and provide some degree of robustness to the frame.
- There may be some useful composite action between beams and floors.
- The weight of the masonry floors may have been over-estimated.
- The live load used to check the structural performance may be unreasonably high (the standard office floor loading in Europe is taken as $5kN/m^2$ when in fact floor loadings actually seldom exceed half this value).

If the conclusion of this assessment is that the structure is inadequate it may be that suitable remedial work can be devised.

Remedial work

The inadequacy of any structure can be remedied in two ways – reducing the loads on the structure, and strengthening the structure. In the example of the 19th-century building the following interventions in the building were proposed and undertaken:

- replacement of the existing concrete/rubble floors with lightweight concrete;
- strengthening of the floors by creating more effective composite action between beams and floors.

Structural appraisal – wrought iron and steel

Despite the differences between the properties of cast iron and wrought iron and steel, the approach to structural appraisal is the same.

The most important issue concerning the reuse of all old buildings is that it is essential that the engineers involved know the methods of construction typical of the building being investigated and have had experience working on the reuse of similar buildings.

For buildings constructed since the late 19th century, it is possible that original drawings and design calculations may exist. Or, if no calculations exist, it will be reasonable to assume that they were done in accordance with the structural design codes and analytical methods that were being devised and adopted as standards and as techniques included in the growing number of textbooks that were being published. Knowledge of such methods will give the modern engineer valuable insight into how the building was designed and constructed.

In the case of steel buildings it can be especially useful to find out which firm made the steel structure – steel-makers' marks were usually rolled into the surface of the steel giving not only its origin but also the grade of steel – its indicative strength.

Compared to timber, cast iron and reinforced concrete, steel has two advantages when reuse is being considered: first, the material can be easily repaired, strengthened or replaced by cutting and welding; and second, this process is the easier because steel is also nearly always hidden from view, so repairs cannot be seen.

Warranty

As with the reuse of building structures using other materials, the structural engineer will effectively take responsibility for the performance of the reused structure.

Reuse of salvaged or reconditioned products and reclaimed materials

Availability (quantity, size and shape)

Cast iron

Being very brittle, cast-iron beams and columns are easily damaged during demolition or deconstruction so care is needed if they are to be reused. Since the casting process allows bespoke sections and sizes to be made easily, such components were made precisely to suit their purpose and location in the original building – the overall dimensions, cross-sections and end fixings. For example, the columns on each floor of a six-storey building would usually have different cross-sections to reflect the increasing load they must carry, progressing from the top to the ground-floor level. Column heights would also vary from storey to storey. This generally makes their reuse less likely.

Cast-iron structural members can be sourced from some architectural salvage firms. However, such firms are more likely to deal in items that are of use for decorative rather than load-bearing purposes. In the UK, the website, www.buildingconservation.com, provides a products and services directory that lists suppliers of reclaimed cast iron, as well as specialists in restoration of historic structures.

Wrought iron

Apart from roof trusses, there is little likelihood of wrought-iron structural components or members being reused. If such a roof structure is dismantled and re-erected it would be preferable to use wrought-iron rivets (rather than steel) to avoid corrosion that can occur where different materials are in contact.

Steel

Since the earliest days of steel construction (1880s), most steel removed from a demolished building finds its way back to the steel-making furnaces where it can be mixed (between 25 and 50 per cent) with virgin steel to make 'new' steel. Only a limited number of steel beams and columns find their way to salvage yards where they are usually sold in small quantities for use in small-scale refurbishment, virtually at the level of local builder. The cost of storage generally mitigates against salvaging steelwork in large quantities.

Many steel beams used since the 1950s have been used with profiled metal decking floors in 'composite construction'. Here studs are welded to the upper flange of the beams and embedded in the concrete topping of the floor. During demolition, the studs, and even the whole beams, would generally be damaged beyond reuse.

Of all the elements in a building, the steel frame is the easiest to imagine being used like the components of a children's construction kit and be almost indefinitely reusable.

Steel beams and column sections are made in standard sections and this makes it easy for a structural engineer to specify structural components (whether reclaimed or new). Likewise, for several decades now, most steelwork has been constructed using a small range of standard connections, mainly of two types with different bending moment capacities – one rigid, the other effectively pin-jointed with no moment capacity. Each type uses bolts of the same size (20mm). The bolts themselves could also usually be reused since they will probably not have been damaged in their first use. This may not be the case with high-strength friction-grip (HSFG), which are usually tightened up to near the limit of their elastic behaviour (Bowman and Betancourt, 1991). The advice of the manufacturers of the bolts should be sought before advocating their reuse.

Supply difficulties are more likely to arise when seeking sections of particular lengths because the ends of the components will have been fashioned to suit a specific connection and also because, even in a very regular steel frame structure, many lengths of beam are used and none of them will correspond to the precise size of the column grid, measured from column centre to column centre. Nevertheless, in principle, usable quantities of such items could be found and cut down, if necessary, to suit a smaller column grid in the new building.

Although this approach is viable in principle, another factor is likely influence the decision to use reclaimed structural steel. Column grids in buildings have to be chosen to suit the dimensions of other building components, especially in the building façade. The manufacturers of windows and façade systems tend to favour certain sizes (e.g. 750mm), which may not suit reclaimed beams that have been cut down in length to remove unwanted end details. Such difficulties can obviously be overcome, but maybe at additional expense. The problem is likely to be easier to overcome if a masonry external envelope is used since this can more easily fit (nearly) any distance between columns.

Consideration should also be given to internal fittings, such as desks and other furniture that need to be fitted between columns.

Structural appraisal

The structural appraisal of individual iron or steel components must follow the principles contained in the procedures outlined at the beginning of this chapter. Generally it is much easier to assess the condition of a component once it has been removed from its original position in a building.

In the case of cast iron, a structural engineer should be chosen who is familiar with buildings of the type proposed for reuse, both in general and, preferably, in the same area of the country, since much construction has its regional idiosyncrasies. A detailed knowledge of the construction material is also essential.

Despite the technical feasibility of reusing structural steel elements it is still seldom undertaken. Storage, cleaning and obtaining suitable quantities are the main hurdles. Nevertheless, various initiatives are being trialled to demonstrate that the hurdles can be overcome (see SCI, 2000: Lazarus, 2002: Wong and Perkins, 2002; case studies in Chapter 2).

Warranty

As with the reuse of entire building structures, the structural engineer will effectively take responsibility for the performance of reused structural components, subject to works being carried out as specified.

Recycled-content building products

As explained in Chapter 3, there is no particular value in the short term in specifying goods with a recycled content of iron or steel.

Structural frame: In situ concrete and precast concrete

Reinforced concrete has been in widespread use since the 1890s and a great many concrete buildings from all periods survive in good condition. Liquid concrete is poured into a mould, usually made of timber 'formwork', inside which has been placed a carefully designed grid of steel rods, bars or cables. Like cast iron, concrete is both weak in tension and brittle. In reinforced concrete, the steel rods are placed so that they carry the internal tension forces making a composite material that can safely carry tension, compression, shear and bending loads. Most reinforced concrete is cast in situ to create bulk form and a monolithic whole. These characteristics have been the main attraction for architects who have used reinforced concrete as a sculptural material, rather as their ancestors used stone.

An alternative strategy for using reinforced concrete is to make individual structural elements resembling those found in steel or timber-framed buildings – columns, beams and floor units or planks. Larger units can also be made, such as components of a portal frame structure, panels used to form load-bearing or non-load-bearing walls, and whole staircases. The individual elements are connected together by mechanical interlocking or bolting together steel fixing plates cast into the ends of the precast concrete elements. The air gaps around such connections are usually filled with a cement grout to provide some additional stability and to afford protection to the steel against corrosion and fire.

Concrete is very durable and good quality; mass concrete still survives from Roman times. Concrete reinforced with steel, however, has a shorter life as water and air gradually penetrate beneath the concrete cover to reach the reinforcement and begin the chemical reaction known as corrosion. Damage to the steel reinforcement is ultimately catastrophic because it is the steel that carries the tension stresses inside the material. If the quality of the concrete is good, and it was well placed in situ, the rate of air and water ingress is very slow and many reinforced concrete buildings are now approaching a century of continuous use after two or three changes of use and minor refurbishment to the concrete elements themselves.

It is in the nature of in situ reinforced concrete that it cannot be dismantled and salvaged for reuse in another building. Precast concrete structures, however, can often be dismantled with little damage, as long as the grouted connections can be undone. Finally, recycled-content concrete can be made in which a proportion of the virgin gravel aggregate or the cement is replaced by recycled materials.

An important factor that will determine the potential for reuse and the use of recycled-content concrete products is whether the concrete will be visible in its finished situation. If so, great care will be needed to ensure repairs match the original material and that consistency is achieved between different batches of new concrete. These considerations are already very familiar to users of concrete in buildings.

Reuse in situ

Structural appraisal

The same general procedures should be followed for structural appraisals that were outlined at the beginning of this chapter. Specific guidance on existing concrete structures is also available (e.g. MacDonald, 2003).

Materials and structural assessment

The strength of reinforced concrete depends on the strength of both the concrete (mainly in compression) and the steel (mainly in tension). The strength of the bond between the concrete and the steel is also important as this enables the concrete and steel to work compositely (together).

The quality, strength and durability of reinforced concrete can become diminished in many ways and due to a variety of causes.

Design of material and structure

- incorrect mix design, especially water–cement ratio;
- poor specification of aggregates;
- poor reinforcement detailing, for example that prevents flow of concrete into moulds;
- inadequate design for creep;
- poor external detailing, for example that leads to water ingress or surface staining.

Quality of materials and construction

- type and quality of cement;
- purity of water used;
- low strength of aggregate;
- weakness of reinforcement, for example due to embrittlement;
- inadequate concrete cover to reinforcement, leading to spalling and/or corrosion;
- chemical actions of additives or contaminants;
- poor distribution of concrete around reinforcement, for example due to impeded flow or lack of vibration;
- entrapment of air and poor compaction during placement;
- poor surface finish, for example due to adhesion to formwork;
- incomplete mixing of ingredients, for example settlement of large aggregate at the bottom of formwork;
- poor bond between concrete and reinforcement, for example due to corroded reinforcement.

Deterioration in service

- actions of carbon dioxide and acidic gases in the atmosphere, for example carbonation;
- actions of chemicals in groundwater;
- spalling caused by frost damage (water ingress followed by freezing);
- biological growth;
- weathering (wind and rain);
- inadequate maintenance and poor repairs;
- relative movement of components due to thermal expansion/contraction;
- settlement and movement of foundations;
- fire damage;

- unforeseen loadings, for example increased floor loadings due to change of use, vibration from traffic, seismic events.

Despite this long list of potential defects to concrete, much concrete is of good and even high quality. Careful investigation can detect the defects and many of them can be rectified or overcome by remedial action enabling the life of a concrete building to be prolonged by many decades.

A thorough visual inspection of the concrete can establish serious problems with the quality of the concrete itself.

The chemical composition of the concrete, its density and its strength can be determined by taking a core sample and undertaking appropriate tests in a materials testing laboratory. The precise geological identification of the aggregate can often give a reliable indication of its strength.

In order to determine the strength of key structural elements it is essential to establish the location and size of the steel reinforcing bars inside the concrete. Several approaches to this challenging problem are possible and, usually, several of them are used in combination:

- Documentary evidence about the original building, including reinforcement drawings for the building, name of the structural engineer and contractor (whose design and construction habits may be known), specifications.
- Identification of a proprietary reinforcement system (for concrete before around 1920 and for building systems in the 1950s and 1960s) with the aid of contemporary catalogues. This will help indicate the likely type, size and location of steel reinforcement.
- Precise dating of the building to establish which design codes would have been used by its original design engineers.
- Removal of concrete cover (in a non-crucial area) to reveal reinforcement; existing damage or spalling can make this an easy task; such damage can easily be repaired.
- Detection of reinforcement using infrasound or X-ray technologies.

In these ways sufficient technical data can be collected to allow the structural engineer to calculate the strength of beams, columns, floors, foundations and shell roofs, and their performance under load.

A guidance document on the protection and repair of concrete structures has recently been produced by the British Standards Institution (BS EN 1504 – *Products and systems for the protection and repair of concrete structures. Definitions, requirements, quality control and evaluation of conformity*); this guidance, published in ten parts over several years, will be adopted throughout Europe and will be of use in any country. Six strategies are identified for addressing defects, damage and decay of concrete buildings and structures, according to the results of the materials and structural assessments:

1 do nothing;
2 reconsider the structural capacity of the building (downgrade building use to suit condition);
3 arrest or reduce further deterioration;
4 repair/remediate decay, and upgrade structure if necessary;
5 reconstruct part of the structure;
6 demolish.

Between 'do nothing and 'demolish' there is likely to be the case where a structure is judged not to be adequate, but suitable remedial action can be identified and undertaken at an acceptable cost to allow the structure to be reused. As in the case of the cast-iron frame discussed above and in Chapter 2, even if a conventional analysis of structural behaviour would seem to indicate that the structure does not satisfy modern design codes, a more careful, bespoke analysis may enable a higher performance to be justified, especially if design loads can be reduced from the normal (high) modern levels.

Remedial work

Many different ways are now available for treating concrete structures and the draft guidance from the British Standards Institution groups them under 11 basic approaches to the protection and repair of concrete:

1 Protection against the ingress of agents (such as water, gases, chemical vapour, other liquids and biological agents) by impregnation, surface

coating, crack bandaging, filling or transferral to joints and membrane application.

2 Moisture control within a specified range of values by impregnation, surface coating, sheltering or overcladding, or electrochemical treatment.
3 Concrete restoration by mortar application/ patch repairs, recasting, sprayed overlays and replacement.
4 Structural strengthening to increase or restore load-bearing capacity by the addition or replacement of reinforcement bars, externally bonded reinforcement, additional concrete, crack/void injections and prestressing.
5 Increasing physical resistance to physical or mechanical attack by overlays or impregnation.
6 Increasing chemical resistance by impregnation, overlays or coatings.
7 Preserving or restoring the passivity of the reinforcement by increasing concrete cover, replacing contaminated or carbonated concrete, re-alkalization or electrochemical chloride extraction.
8 Increasing the resistivity of the concrete by limiting moisture-using surface treatments, coatings or sheltering.
9 Cathodic control to prevent anodic reaction (of the steel reinforcement) by surface coating or saturation.
10 Cathodic protection (of the steel reinforcement) against further corrosion.
11 Anodic control to prevent anodic reaction by painting the reinforcement with active pigments or barrier coatings and inhibitors.

Although these many remediation techniques need to be undertaken by specialists in their field, they fall into two types: first, those that influence the load-carrying capacity of the reinforced concrete; and second, those that lengthen the life of the structure by arresting decay of the reinforced concrete itself. The former type of remediations need to follow the recommendations of a general structural engineer who would be responsible for setting the precise aims of the remediation in terms of achieving resistance to certain specified loads.

While details of the many methods of strengthening concrete structures are beyond the scope of this book, it is worth illustrating the use of carbon fibres to provide additional, external reinforcement to augment the contribution of the internal steel reinforcement. All loads need to be removed from the structural member (even the self-weight of the structure, by jacking off the floor below) before the sheet of carbon fibres is bonded to the underside of the beam with an epoxy resin. After hardening, the loads can be re-imposed and carbon fibres will carry a predetermined proportion of the bending stresses in the beam.

Warranty

The structural engineer responsible for the appraisal and any remedial work will be able to warranty the structural performance.

Reuse of salvaged or reconditioned products and reclaimed materials

While there may well be an exceptional case, it should generally be assumed that in situ concrete cannot be reclaimed for use in another location.

In principle, many precast concrete elements could be removed from a building, refurbished if necessary, and reused. This could apply particularly to:

- columns, beams and portal frames;
- floor planks made of ordinary or pre-stressed reinforced concrete;
- staircases (usually in units of a single straight flight);
- panels forming internal partitions or external walls (cladding panels are discussed in Chapter 6);
- blocks forming part of a proprietary flooring system.

The success of such an operation will depend crucially on two factors:

1 the condition of the reinforced concrete itself;
2 the ease with which the components can be separated.

The condition of the reinforced concrete can be established in the same ways as outlined above for a concrete structure remaining in situ. According to the assessment of the materials, the components could be protected or repaired in the same manner as an in situ structure.

A great deal of structural precast concrete comes in the form of a proprietary system. Different proprietary systems have different methods of fixing the components together and the most practical opportunity for reuse would be to salvage all the components from a building and keep them together as a set. This would mean that the methods of fixing components together would be compatible.

The different methods of fixing components together have different degrees of reversibility. Many systems depend on steel bolts to join plates embedded in the beams and columns. While these can be easily unbolted, it may be necessary to remove grouting material used to embed the connections to protect them from corrosion and fire. Other systems rely on a more substantial joint cast in situ – individual components are made with reinforcing bars protruding and a substantial amount of concrete and maybe some additional reinforcement is used to form a frame that more nearly resembles one cast in situ. Such a system will be much more difficult, if not impossible, to disassemble without serious damage.

In general the capacity for precast concrete units to be reused depends on whether the firm that devised the system considered its demolition or deconstruction. In the future, it is likely that deconstruction will be given greater consideration in the design of structures, and precast concrete will be a prime material to reap the benefits of reuse.

As with the reuse of all materials, it will be made easier if accurate information about the product and its materials are made available to the potential user. Trials are already under way in several countries in the use of 'e-tagging' where a microchip containing full structural and material information is fixed to the precast concrete product.

Recycled-content building products

There are four types of opportunity to use recycled materials in concrete construction:

- the replacement of cement by other materials in in situ concrete;
- the replacement of virgin gravel as the aggregate in in situ concrete;
- the use of concrete blocks and other items made from concrete with recycled content (namely the previous two options);
- various ancillary products involved in concrete construction made from recycled-content plastics.

Cement-replacement materials for in situ concrete

The manufacture of cement requires burning limestone and other materials at very high temperatures and, for this reason alone, any means of reducing the amount of cement used in concrete will bring environmental benefits. There is a double benefit when the material used as a cement replacement is itself a waste material that would otherwise have to be disposed of in a landfill site. The use of cement replacement materials was developed in the early 20th century mainly for civil engineering uses. Its use has spread to building applications in more recent times.

Pulverized fuel ash

Pulverized fuel ash is the ash carried out of a furnace in the waste gases following the combustion of pulverized coal in coal-fired power stations. It is a fine powder, comprising mainly silicon, iron, aluminium and varying amounts of carbon in the form of minute glass spheres, that resembles Portland cement in both colour and texture. PFA is a pozzolanic (hydraulic cement) material that can be blended with Portland cement and will harden during its reaction with water, even in the presence of excessive proportions of water. It can be added to the concrete mix either as a separate ingredient or as a pre-blended mixture with Portland cement. Proportions vary according to the type of PFA and desired concrete mix design, but typically constitute 25–40 per cent of the PFA/Portland cement mix.

PFA has been in use for over a century and many countries have standards that govern its use in making concrete. Apart from the environmental benefits, concrete made with PFA can bring a number of additional benefits compared to concrete made only with Portland cement, including reduced cost, improved workability during the making of concrete, improved long-term strengths and durability, and reduced shrinkage and creep.

Ground granulated blast-furnace slag

Ground granulated blast-furnace slag is produced as a by-product of iron-making in the form of glassy granules, and consists of compounds of the oxides of calcium and magnesium together with silica and alumina. The form of GGBS suitable as a cement-replacement product is produced when the slag is rapidly quenched in water (rather than slower air-cooled quenching). It has been used as a cement replacement for over a century and many countries have standards that govern its use in making concrete. It has been available as a separate ingredient to add to the concrete mix since the 1960s. Apart from the environmental benefits, concrete made with GGBS can bring a number of additional benefits compared to concrete made only with Portland cement, including lower heat of hydration (reduced risks of cracking), lower permeability to water, improved durability in aggressive environments and higher long-term strength. GGBS is popular with both architects and engineers, as it produces a light coloured concrete with increased resistance to alkali-silica reaction (a cause of concrete decay). One disadvantage is that concrete made with GGBS gains its strength more slowly than normal concrete and so formwork must be left in place for longer before it is removed.

Both the European (BS EN 206-1) and UK (BS 8500-1, BS 8500-2) Concrete Standards incorporate the use of cements that include blends of Portland cement with a wide variety of other main constituents, including PFA and GGBS. In addition, the UK Concrete Standard includes procedures for the use of combinations of Portland cement with either PFA, GGBS or limestone added at the concrete mixer, where the PFA, GGBS or limestone is accepted as part of the cement.

Aggregate-replacement materials for in situ concrete

Naturally occurring aggregates are made from rock cut from quarries or from gravel dredged from the sea-bed. The precise geological properties of aggregate are particularly important since these determine the appearance of concrete. When a particular colour of concrete is required, stone from a named quarry may well be specified.

Enormous quantities of aggregate are used in building construction, although this is only a fraction of the quantities used in civil engineering construction. Using waste materials to replace natural materials leads to a reduced demand for non-renewable resources as well as reduced environmental impact from quarrying and trawling activities, including the impact from vehicles transporting the virgin materials (mainly) from countryside or seaside locations to inner cities. Other benefits of using recycled materials instead of new aggregate and cement are:

- less materials are sent to landfill for disposal;
- the recycled materials will usually be cheaper than new materials;
- using recycled materials to replace natural aggregates will avoid having to pay taxes on the use of extracted primary materials, for example the 'aggregates tax' in Britain;
- the amount of embodied energy/carbon in the building will be lower than an equivalent building constructed of new materials;
- recycled materials are likely to be generated locally, hence reducing transport requirements.

It is important to ensure that recycled materials are used according to their quality. High-quality materials should be used only where high-performance materials are required (e.g. columns) and low-grade materials should be used in low-grade applications (e.g. on site roads and in foundations).

It will not always be possible to use recycled materials to meet all a building's concrete requirements. Where fresh aggregate and cement are needed to fill a shortfall of recycled materials, these should be used where high-performance material is required, and the recycled materials should be used for lower-grade applications.

A great many materials that would otherwise be waste and sent to landfill sites can be used as replacements, though since they affect the colour of the final concrete, they cannot be used indiscriminately. As always with concrete, it will be necessary to make up trial mixes to test for colour as well as strength. The materials used as aggregate replacements include:

- crushed concrete from demolition (only if clean and of consistent quality and colour);
- crushed bricks from demolition (only if clean and of consistent quality and colour);
- slate and other natural stone waste (in 'reconstituted stone');
- pulverized fuel ash (for lightweight concrete);
- crushed, post-consumer glass;
- reclaimed railway ballast (only if clean and of consistent quality and colour);
- reclaimed foundry sand (only if clean and of consistent quality and colour);
- china-clay sand.

Of these materials only a few are likely to be of a quality suitable for use in new concrete used in building construction.

PFA *and* GGBS

Both materials can be used to make lightweight concretes for in situ use and also for making concrete blocks. Lightweight concrete made with PFA can be used for lightweight structural concrete, for example in the upper floors of buildings to reduce dead loads, roof and floor screeds, surfaces for playgrounds and sports playing areas.

Crushed glass

Successful trials have been completed in using finely-crushed glass as a 'micro-filler' in concrete. The product has increased workability and flow properties that reduce the need to add plasticizers. It also has improved long-term strength, durability and resistance to frost. It has been used for in situ concrete floors and walls, concrete pipes and precast elements for constructing floors and walls. These benefits relate to crushed glass where it is ground to a fineness similar to that of Portland cement. Where larger particle sizes are used there is a risk of a damaging alkali–silica reaction, a deleterious expansive reaction, normally between the alkalis from cement with particular forms of silica.

Recycled concrete aggregates for in situ concrete

Currently the greatest interest in using recycled materials for in situ concrete is the use of crushed concrete as an aggregate replacement. One reason for this is the inclusion of using recycled demolition material in environmental assessment tools such as BREEAM, EcoHomes and LEED (see Appendix B). Credits or points can be gained by using recycled materials. Planning guidance from local planning authorities is also emerging on sustainable construction, as is best practice guidance on demolition. These encourage the use of RCA in building construction.

RCA is suitable as a replacement for coarse aggregate. It is not suitable as a replacement for fine aggregate (sand) for two reasons: First, the larger surface area of secondary material leads to unacceptable leaching of salts from the crushed concrete; and second, it is difficult (time-consuming and expensive) to achieve a suitable range of particle sizes as the crushing process creates too great a proportion of fine particles (dust).

The use of RCA is generally considered for main load-bearing structural elements (columns, walls and beams) and for lower performance duties such as floors and basements. RCA is generally considered unsuitable for use where concrete will be left fair-faced and exposed due to the difficulty of achieving consistent colour and texture.

Before discussing such uses in more detail, it is worth reflecting on whether to use crushed concrete as RCA or not and, if so, what to use it for. There are, in fact, a great many uses for large quantities of crushed concrete in civil engineering applications where a lower grade of material is often acceptable, such as hardcore, load-bearing fill and road construction. Indeed, demand for such material exceeds the supply of suitable materials from demolition waste. If virgin aggregate needs to be quarried to meet this demand, it can be argued that it would be more sensible to use all demolition waste for low-grade uses and use natural, high-quality aggregate for purposes where high structural performance is needed or where the concrete is visible. It is also worth reflecting that the cost of concrete for use in buildings made with RCA is typically 5–10 per cent more expensive than concrete made with natural aggregates.

The balance of supply and demand for crushed concrete for use as a replacement for natural aggregate will vary from country to country, place to place and with time, so a careful assessment will need to be made for each project. It may therefore be that the balance is tilted not by environmental considerations, but by market conditions and political objectives linked to increasing the recycling of materials.

The one time when using RCA is likely to be the preferred choice is when a building is being demolished on the site where a new building is to be constructed and the crushing and making of new concrete can be undertaken on the site. In such a case the precise origin and quality of the source material will be known and the saving in transport costs is likely to be significant. If this is to be achieved, care will be needed to set up the demolition contract in such a way that the main contractor for the new construction will be happy to accept the crushed concrete left on the site by the demolition contractor.

Generally speaking, aggregate made from crushed concrete will need to meet the appropriate standards that define the specification of natural aggregates. Work is underway in several countries to develop specifications dedicated to RCA, but none are yet in force. The following example of UK specifications is indicative of the situation is many other countries (Box 5.1).

There is still only sporadic information about the suitability of crushed concrete as a recycled aggregate, mainly because the properties will vary with the source of the material. De Vries (1993) has made the following comparison between the properties of concrete made with new ingredients and with crushed-concrete aggregate:

- Compressive strength: replacement of up to 20 per cent of coarse aggregate by crushed concrete has little effect. Replacement of 100 per cent of aggregate by crushed concrete will reduce compressive strength by 10–20 per cent.
- Stiffness: replacement of up to 20 per cent of coarse aggregate by crushed concrete has little effect. Replacement of 100 per cent of aggregate by crushed concrete will require structural members to be around 10 per cent thicker or deeper.
- Durability: for the same strength concrete and mix design, the replacement of natural aggregate by crushed concrete has little effect.
- Workability: replacement of up to 20 per cent of coarse aggregate by crushed concrete has little effect. Replacement of larger proportions significantly reduces workability and more water must be included in the mix (this is due to more irregular shapes of the aggregate, the greater absorption by crushed concrete and the presence of unhydrated cement).

Box 5.1 Specifications for RCA in the UK

The UK Concrete Standard BS 8500-2, *Concrete – Complementary British Standard to BS EN 206-1 Part 2: Specification for constituent materials and concrete*, sets out the conditions under which RCA can be used in concrete. Essentially the requirements limit the use of RCA to a maximum concrete strength class C40/50 and where the exposure conditions exclude chloride ingress, high water saturation freezing and thawing, or chemical attack including that from aggressive ground.

Procedures for ensuring the quality of recycled aggregate are described in *The Quality Protocol for the Production of Aggregates from Inert Waste* (Waste and Resources Action Programme, June 2004).

A UK government consultation paper *Mineral Planning Guideline 6 and Aggregates Provision in England 2001–2016*, published in 1996, proposes that the aggregate supply for 1996–2006 should come from 12 per cent recycled and secondary sources, as opposed to an estimated 10 per cent in 1989. The equivalent document for the period 2001–2016 is currently available in draft form and proposes a much larger figure, with up to 25 per cent of aggregates being supplied from recycled and secondary sources.

Source: www.odpm.gov.uk/stellent/groups/odpm_planning/documents/source/endnotes

This work and more recent research in a number of countries supports the general guidance that, for concrete strengths up to 50N/mm^2, up to 20 per cent of natural coarse aggregate can be replaced by clean, crushed concrete with no adverse effects on the resulting concrete. Outside these limits, specific testing should be undertaken (de Vries, 1993).

Information about the use of RCA is growing steadily, though in terms of quantities, its use in civil engineering projects dominates its use in building construction. Many universities and industry research organizations are conducting research into the properties of concrete made with RCA, and professional engineering journals now regularly report the outcomes of such research.

The supply of RCA is also becoming easier to ensure. Many countries now have internet-based materials exchanges that list suppliers of RCA, specific quantities of available material as well as potential users with specific needs for RCA. In the UK, the Aggregates Advisory Service can provide a list of authorized crushing plant operators and producers of recycled aggregates.

General information about concrete can be obtained from concrete industry trade associations and various organizations promoting recycling as an important means of reducing the quantities of waste sent to landfill, such as the Concrete Centre (www.concretecentre.com) and AggRegain (www.aggregain.co.uk) (in the UK).

Precast concrete blocks and other products made with recycled-content concrete

Concrete made with PFA can be used to make a variety of precast concrete products used in the construction of buildings, including:

- concrete roof tiles;
- concrete pipes;
- concrete blocks;
- concrete lintels;
- kerb edgings;
- fence posts;
- flags and paving stones.

There are numerous internet-based directories throughout the world that list suppliers of materials. For example:

- The California Integrated Waste Management Board maintains a Recycled Content Product Directory (www.ciwmb.ca.gov/RCP/Default.asp), which lists thousands of recycled products and provides information on the companies that reprocess, manufacture and/or distribute these products.
- The AggRegain website (www.aggregain.co.uk) contains a directory of more than 250 suppliers of recycled and secondary aggregate products at 350 locations throughout England. The supplier database, which is part of the UK Recycled Products Guide is operated by Waste Watch, and is regularly updated.

Ancillary items in concrete construction

Finally it is worth mentioning some opportunities to specify recycled-content products among the ancillary goods involved in concrete construction:

- One way of reducing the amount of material used in in situ concrete is to create large voids inside. This method also reduces the weight of the concrete structure, which means smaller loads on foundations and the possibility of smaller foundations or piles. The formwork for such voids can be made using a lightweight mixture of recycled polystyrene (around 80 per cent by volume) and cement. Its very low density means it is easy to handle and blocks of the material can easily be cut to shape and glued together on site as required. The material has excellent insulation properties, is fire resistant and is also resistant to termites.
- Steel reinforcement needs to be precisely located inside the timber formwork, and particularly needs to be a minimum distance from the inner surface of the mould. The plastic spacers that clip onto the reinforcement to maintain this position can be made from 100 per cent recycled-content plastics.
- Plastic waterproof membranes used during concrete construction, for example to keep rain off or protect concrete from frost, can also be made from 100 per cent recycled material.
- Large slabs of concrete are usually cast in small sections to allow provision for thermal expansion. The joints between sections of slab must be made of a compressible material that can accommodate the movement. One option for the filler material for such joints is made from 100 per cent post-consumer paper or post-industrial wood waste.

Case Study: Wessex Water

The new operations centre for Wessex Water was designed to be an exemplar building in green design and construction. Apart from very low energy use, and achieving an 'excellent' rating using BREEAM, a key feature of the construction was to use a high proportion of recycled aggregates for the in situ concrete.

The viability was dependent on finding a clean, uniform and local source of suitable concrete for crushing and reusing. The source had to be clean and uniform to minimize contamination and the need for separating different qualities of material, and had to be local to minimize transport requirements and associated cost and environmental impacts. A number of risks were identified, including running out of material, plant breakdown and a lack of a competitive market to negotiate supply. The concrete supplier was finally able to provide assurances that they had sufficient material of a suitable quality when their supplier of aggregates obtained 40,000 pre-stressed concrete railway sleepers that were at the end of their useful life and due to be disposed of. To further minimize the risks, the proportion of coarse aggregate was limited to 40 per cent. Also, the use of recycled aggregate was limited to foundations, the basement and columns. The exposed concrete soffits in the building, which perform a vital role in providing thermal mass and storing heat are of precast concrete made with virgin aggregate.

The railway sleepers provided a clean source that could be delivered to the works by rail, thereby reducing impacts from transport. They were crushed and aggregate in the range 5–20mm was used in the concrete. Material less than 5mm was separated and used as fill elsewhere on site. The aggregates were tested for quality, consistency and performance, and found to be similar in performance and appearance to the local Mendip carboniferous limestone, and conformed to the appropriate UK standards for concrete aggregate. Batching of the concrete was found to be very similar to concrete using local limestone. The concrete contractor found no difference in colour, texture or workability between regular concrete and concrete made from a mix of recycled concrete and natural aggregate. Further to this, there were no problems in handling, pumping, vibrating or finishing the concrete.

Some problems, as well as additional costs, were encountered but were handled appropriately by the project team:

- There were additional start-up costs due to mix design and testing.
- Additional storage facilities were required, along with a loading shovel for the recycled aggregates and some reorganization of plant, quality-control procedures and testing. These too resulted in additional costs. However, the concrete supplier expects that these costs would reduce with scale and that the future cost of concrete made from recycled aggregates could be comparable to regular concrete.
- Some disruption was caused to other local customers of the concrete supplier, as some contracts had to be supplied from other plants.

Overall, the additional capital cost of using crushed sleepers represented approximately 5–6 per cent of the cost of the concrete placed in situ. However, a number of benefits were achieved, including the following:

- The environmental impacts of transport were reduced by delivering the sleepers by rail. It also prevented the need to dispose of the sleepers to landfill.
- The process identified a new source of material for recycled aggregates. It also helped in the formulation of a quality-control system for recycled aggregates and a cost-effective, no-waste method of processing the sleepers. These all have positive implications for future use of recycled aggregates.

There were a number of risks associated with the use of recycled aggregates in concrete, however, these were managed through a combination of foresight, effective communication throughout the supply chain, sound technical advice, and setting a limit of 40 per cent on the proportion of recycled aggregates.

The client continues to be committed to becoming a more sustainable operation and has judged that the small additional costs associated with the use of recycled aggregates were more than balanced by the value associated with long-term sustainability benefits.

Location: Near Bath, UK
Date: 1997–2000
Client: Wessex Water plc
Architect: Bennetts Associates
Services and structural engineer: Buro Happold
Adviser on recycled aggregate: Building Research Establishment
Concrete contractor: Byrne Brothers
Refs.: Anon, 2001.

Floors in the structural frame

Until the late 18th century, floors in nearly all buildings were made using timber floor beams supporting wooden planks. From that time there developed the first multi-storey commercial buildings for warehouses and factories housing many industries and goods, but notably the cotton and wool industries. The machinery in these buildings, the steam engines that powered them, and the heating systems in the building all increased the risk of fire and many buildings and lives were lost in terrible fires. These events soon led to the development of so-called 'fireproof construction' that mainly consisted of replacing timber beams and columns by cast iron and, later wrought iron and steel, and replacing timber floors by a variety of incombustible alternatives.

Throughout much of the 19th century, two types of floor structure predominated: the brick jack arch spanning parallel iron or steel beams, topped with concrete and a floor of stone flags, and a series of primary and secondary beams supporting square stone flags.

By the late 19th century most floors in industrial and other large buildings with steel or reinforced-concrete frames were made of concrete – either cast in situ with some rudimentary iron or steel reinforcement, or made from precast blocks of clay or concrete, usually with air voids to reduce their weight and increase fire resistance. A huge number of very similar proprietary systems were patented between the 1860s and 1920s, and many similar systems using precast units are still in use today.

In steel-frame buildings profiled steel decking floors were first used in the US from the 1910s. The corrugated steel sheet is topped with a layer of concrete both to provide a robust surface to the floor and to ensure adequate fire resistance between storeys. This concrete will be some 100–150mm thick with nominal reinforcement to prevent shrinkage cracks. Since the 1950s various means have been used to develop composite action between the steel frame and the steel sheet and concrete floor to reduce the thickness of concrete needed to achieve the necessary strength and stiffness. The bond between steel beams and concrete is typically achieved by welding studs to the top flange of the steel beams that protrude through the metal decking into the concrete. This method of construction is clearly very difficult to dismantle without damage.

In situ reinforced concrete buildings usually have floor structures that are cast integrally with the frame. Depending on the span of the floors, and whether the soffit is exposed or chosen specifically to provide space for building services, the floor may be a flat slab or smaller areas of slab spanning beams of reinforced concrete.

Precast-concrete frame buildings usually have floors comprising compatible precast planks or hollow blocks spanning precast beams. As with the frame itself, different proprietary precast systems require different types of grouting or finishing with an in situ concrete topping that renders the precast system more or less easy to dismantle.

Reuse in situ

All floors can be reused in situ subject to a demonstration that the beams and other components can carry the load likely to be imposed. The process of assessing the capacity of any floor is the same as for other parts of the structure, as described earlier in this chapter.

In the event that a beam is found to be deflecting too much, especially in the case of a timber floor beam, it is nearly always possible to stiffen the existing beam or, if there is damage or irreversible decay, replace it.

Reuse of salvaged or reconditioned products and reclaimed materials

Timber floor beams and floorboards, salvaged from another building, can easily be reused, subject to the same assessments of the timber described above for the timber frame.

As with the cast-iron structure of 19th-century industrial buildings, it is unlikely that anyone would want to use an original brick jack-arch floor from another building, except in the case that components from one part of a building being refurbished might be cannibalized to repair another part of the same building. This type of floor structure is relatively easy to dismantle and rebuild.

Neither reinforced-concrete floor structures nor profiled metal decking floors can realistically be removed from a building and reused elsewhere.

The components of floor structures made from a proprietary precast-concrete system may be reusable if they are only lightly grouted together to give structural integrity to the building. In principle it would be possible to reuse such components, though at present it is very unlikely that a salvage firm would reclaim them in the hope that someone would buy them. The most likely opportunity for reuse is when the structure of a building, including the floor structure, has been identified as a candidate for reuse prior to its demolition.

Recycled-content building products

The main opportunity for using recycled-content materials in floor structures is the use of concrete made with cement or aggregate replacements. The same guidance applies as to using these materials for other structural concrete, given in the previous section. It is worth giving special mention to using lightweight concrete made with an aggregate replacement of PFA or expanded blast-furnace slags. These are both suitable for floors and, if there is a need to reduce the dead weight of an old floor, the existing concrete or rubble can be removed and replaced by a lightweight concrete.

Floor structures made of precast concrete units offer the opportunity to use blocks or planks made from concrete with recycled content, either cement or aggregate replacement. Such products can be found by searching in the online databases for recycled-content building products.

6 Design Guidance: Building Envelope

Reclamation, reuse and recycling in the building envelope

Several words are used to describe the visible, external envelope of a building – façade, envelope, wall, cladding, roofing. Each tends to reflect who is using the word and why. There is no consistent set of definitions that applies to all usage in all (English-speaking) countries. An architect is most likely to be concerned with visual appearance; a structural engineer with whether the visible materials are load-bearing or not; a building services engineer with how the complete sandwich of different layers act as selective filters to keep water and water vapour out, keep heat in and allow light into or out of a building. Façade engineers would expect to embrace all these aspects. When considering reuse and recycling, it is the materials used in the envelope that are of primary concern, and how these are used to achieve the desired performance.

In masonry buildings the load-bearing walls serve both as part of the structure and the building envelope. In modern buildings the function of structure and envelope are largely separated into independent building systems – usually a steel or reinforced concrete structural frame and a cladding system that is supported by the building frame. In either case, the building envelope has to perform a number of distinct functions and these must all be considered when reusing a façade or roof, or using reclaimed or recycled-content materials, just as when using new materials and components.

The value of reuse

Being the most visible part of a building, the external envelope contributes more than any other part to people's impression of a building and, especially, to its architectural style. The nature of the façade and roof of a building determine its relationship to adjacent buildings and the character of the area around the building. Vernacular architecture is strongly characterized by the materials used in the façade and roof and until the mid-19th century would nearly always have been constructed using locally available materials, whether natural (timber or stone) or man-made (bricks and tiles made using local clay).

The appearance of the façade and roof of a building is given great consideration when local authorities are consulted to give planning permission. Most local authorities provide guidance for building designers on suitable and appropriate materials and types of construction. In most villages, towns and cities there are areas designated as conservation areas or areas of cultural or heritage importance. In such areas both construction materials and the appearance of buildings may be very tightly prescribed.

For many buildings, therefore, an important reason for retaining old roofs and façades, or using reclaimed materials in roofs and façades, might be to ensure that planning permission is awarded.

Another major influence on the decision to reuse components or use reclaimed materials is the sheer mass of materials in the roof and building façade that is second only to the quantities of material used in the load-bearing structure of a building. In load-bearing masonry buildings, the façade and the structure are one and the same, and their reuse will divert a large quantity of material from being sent to landfill sites. This can save money by avoiding landfill taxes.

Although it is unlikely that a detailed environmental impact analysis will be undertaken when choosing whether to use reclaimed materials, common sense should prevail when they are being sourced – the adverse environmental impact of road transport on air quality is not insignificant. There is thus likely to be little or no overall environmental benefit to using a small batch of reclaimed materials if they have to be transported many hundreds of kilometres from source to building site.

Opportunity and availability

The life expectancy of building façades is generally lower than the structural elements of a building that are usually protected from the external environment. This is partly because they are more exposed to the elements and various agents of decay. While most types of building façade can be refurbished, building owners may feel pressure to make more radical changes, including replacement, in order to keep up with changing fashions in the appearance of buildings.

Masonry façade retention

Most small buildings and many larger ones, especially if built before 1900 or so, have load-bearing external walls that form the façade or building envelope. In recent decades there has been a growing concern for preserving a nation's built heritage while wanting to upgrade the building interiors to modern standards. Going well beyond the normal procedures for refurbishing old buildings, the technique has developed for retaining the visible façade of a valued building.

The retention of the façade is of an existing building consists of four main stages:

1. Assessing the façade to establish its construction and how it is linked to the rest of the building where walls and floors meet it.
2. Designing the new building structure to fit the retained façade, both with respect to geometry and the mechanical link between old façade and new structure (especially the possibility of relative movement between the two).
3. Devising and constructing a means of supporting the façade prior to removing of the original means of support (such temporary works may need to withstand considerable wind loads).
4. Constructing the new building structure and transferring to it the role of supporting the façade.

As with the design and construction of any building envelope, the performance of the old façade as part of the new building envelope will require careful consideration of the various issues outlined above.

The retention of masonry façades is now a well-established practice and further details are beyond the scope of this book. Excellent guidance for structural engineers and building contractors, including case studies, is available (Bussell et al, 2003) and so is not discussed further here (see case study below).

Reclaimed materials and products

The methods of construction used for many types of roof and building façade, both traditional and modern, are such that the components can be removed during demolition with little or no damage and reclaimed to make them suitable for reuse. This is especially true of roofing tiles and slates, and brick and stone used in façades for which there is nowadays a buoyant market. Architectural salvage firms tend to specialize in certain sectors of the market. The materials most likely to be salvaged during demolition are those that are no longer made, such as old bricks and roof tiles, and those that can easily be salvaged for reuse, such as bricks laid using lime mortar rather than modern, high-strength cement mortars.

Modern cladding systems for building façades comprise high-value elements such as granite or less costly metal sheet. Most of these could, in principle, be carefully removed from buildings during demolition and reused. In practice, relatively little is reclaimed at present and a change in the attitudes of building owners, designers, building firms and demolition contractors will be needed before this situation changes. Such changes are, however, possible – only 30 years or so ago hardly any roofing tiles or bricks were recycled; today there is a thriving market.

Recycled-content building products

Apart from the use of reclaimed building materials such as tiles and bricks, a growing number of products for use in constructing the building envelope are available. These incorporate various 'waste' materials such as polymers and crushed ceramic materials used as aggregates for concrete products.

Costs

The costs of reclaimed products and materials for use in the building envelope, compared to their new equivalents, vary considerably according to the items and the abundance of their supply. In particular, costs will be difficult to predict at the outset of the project. If products to be reclaimed and reused are identified prior to the demolition of the building in which they were first used, prices will need to be negotiated with the current building owner or the demolition contractors. Reclaimed bricks and roofing tiles, on the other hand, can be viewed prior to purchase in an architectural salvage yard.

As with all building elements with a finite service life, with intermittent maintenance requirements and the eventual need for replacement, commercial decisions about the building envelope need to be made on the basis of *whole life costing*. This is especially important in the case of the building envelope since its performance, particularly its thermal performance, has a direct and very significant influence on a building's running costs. The viability of using reclaimed materials and components in the building envelope, including the capital costs of providing increases in thermal performance, will need to be tested carefully against the savings achieved by reducing energy costs and any cost benefits of prolonging the life of an existing envelope or using reclaimed products and materials.

Design issues

When designing a building façade to incorporate reused or reclaimed components or materials, designers will need to address the same issues as for new materials and components. The main issues include:

- Appearance – architectural details of all elevations, including materials and colours, usually need to be agreed before planning permission can be granted. It is unlikely that reclaimed materials that are expected to be used will have been identified at this stage. In this case there will need to be discussions and agreement with appropriate representatives of the planning authority before their use can be finally approved.
- Structural requirements – the components of the envelope need to be supported by the structure of the building; the envelope must also carry wind loads on the building to the parts of the load-bearing structure designed to carry wind loads. The envelope will also need to be able to withstand suction loads arising from low air pressure on the lee of the building (in strong winds such forces can lift clay tiles and even suck windows out!).
- Ingress of water – the envelope (or a sample of it) may need to be tested to verify its ability to exclude water.
- Airtightness – the envelope of the building must be sufficiently airtight. Building regulations in some countries require exfiltration tests on the building envelope to verify its airtightness. This is already the case for some buildings in the UK and will be extended to cover housing from 2006.
- Thermal performance – the envelope of the building as a whole must provide adequate insulation, as laid down in building regulations.
- Solar performance – the envelope of the building as a whole must provide adequate solar performance, as laid down in building regulations.
- Condensation – the airtightness affects the ingress and egress of water vapour that can lead to condensation within the thickness of the wall or cladding. To ensure no condensation occurs, the temperature and vapour pressure across the thickness of the wall or cladding must be calculated.
- Acoustic performance – the envelope of the building as a whole must provide adequate sound insulation, as laid down in building regulations. This includes sound transmission between outside and inside of the building as well as transmission between adjacent rooms. Tests may be required to verify performance.
- Fire resistance – the envelope will need to achieve the appropriate fire rating.
- Durability – the external envelope of a building will need to be sufficiently durable, secure and vandal-proof.

A major consideration when designing building façades is the interaction between the envelope and the structure of the building. Because they are constructed of different materials and subject to different conditions, they behave differently with respect to movement caused by temperature change and loads. The physical connections between envelope and structure need to be designed to accommodate any relative movements.

Similarly, the components that make up the building envelope and the structure are of different sizes and constituent components are available in different sizes. When considering the use of reclaimed elements in the building façade, care is needed to match the grid of the building structure (column spacing and floor-to-floor height) to the sizes of the façade elements.

Performance specification and warranty

The construction of the building envelope used to be entirely prescriptive, with materials and dimensions precisely specified, and this approach may still be used for some buildings.

Nowadays, however, the building envelope is usually defined and specified in terms of the overall performance under the various headings listed in the previous section, for example thermal and acoustic performance. This allows the designers a certain degree of freedom when deciding how the required performance will be achieved and there is, in principle, no discrimination against using reclaimed materials. The minimum performance levels are defined in various design codes of practice and building regulations. They apply equally to a new building envelope and one made with reclaimed materials. This approach generally requires testing of the finished construction to verify compliance with the performance specifications.

Generally speaking, the performance of a building envelope can be predicted using engineer's design calculations that serve as evidence of being satisfactory. Where calculations may not be reliable, such as when a façade is made in an unusual way, which may be the case for one made with reclaimed materials and components, the actual performance of the building envelope may need to be proved by tests, either on a full-size prototype or the completed building. These are most likely to include tests on watertightness and airtightness, acoustic and thermal performance.

The required level of performance of the building envelope is gradually increasing in most countries in order to reduce the energy needed to provide the high level of comfort that people now demand of the inside of buildings. One significant consequence of this development is that components and types of construction used in the past may not be able to achieve the level of performance required today by new building regulations.

Cladding systems

A cladding system is a type of skin or envelope to a building that is supported either directly from the structural frame or from a sub-structure that is, in turn, attached to the main building frame.

Various types of cladding system have been developed since the first use of structural frames of iron or steel for buildings in the 1860s. Such frames carry their loads independently of the external and internal walls. In many early examples, especially in Chicago and New York, a full-height masonry façade was built in front of the structural frame and attached to it to ensure its stability. Such a wall had only to support its own weight and so could be much thinner than one that had also to carry all the floor loads too. Further economy was achieved when the façade for each storey of a building was supported from an edge beam; the thickness of such a façade could thus be reduced to that needed for a single-storey building. Cladding systems today are based on this same principle. The cladding is supported at each floor, either from an edge beam or from a separate, lightweight structural grid attached to the floors. A subsidiary function of all cladding systems and their supporting structure is to convey the wind loads impinging on the façade back to the main structure of the building and to the shear walls that convey the loads down to the foundations.

There are two main types of such cladding system:

1 Unitized, panel or strong-back systems consist of many identical, pre-assembled panels fixed directly to the building. They typically span the full storey height and incorporate the glazed areas of the façade. The panels are typically

made of several materials bonded or joined together:
- sheet metal (often aluminium or stainless steel, pressed to form a suitable shape);
- stone (either solid or a face bonded to a reinforced panel);
- precast, reinforced concrete (including 'reconstituted stone');
- glass-reinforced cement (GRC);
- glass-reinforced polymer (GRP);
- laminated or toughened glass;
- fixing brackets, usually of stainless steel.

2 In site-assembled or stick systems, the façade is partly constructed in situ by fixing a series of vertical mullions and horizontal transoms to the building (the 'sticks') that support the individual panels of the façade, which might be opaque (e.g. granite slabs) or transparent (sheets of toughened glass). Such systems contain high-value materials and fixings and are rather labour-intensive to install and dismantle.

Most cladding systems are designed and made as bespoke items for a particular building and make a significant contribution to the building's identity. Nevertheless, there is some standardization in such systems, including overall sizes (e.g. multiples and submultiples of 1500mm) and methods of fixing panels to their supporting frame.

Less costly external façades to buildings can be provided using profiled metal sheeting. This finds widespread use in industrial units, the envelopes of which do not need to provide the same degree of thermal and acoustic insulation as buildings used to house people. Alternatively, profiled metal sheet can be used as rain-screen cladding to provide an attractive visible exterior to a low-cost brick, block or concrete external wall that provides the necessary thermal and acoustic performance.

Timber boards can be used to provide a rain-screen cladding. Timber in very small pieces (shingles) can be used like overlapping roof tiles to form a rain screen.

Reuse in situ

Both strong-back and stick cladding systems can be refurbished in situ to some degree. Concrete and stone exposed to the elements usually acquire some colouration that can enhance the façade's appearance. However, in some atmospheres and if water run-off is not well managed, staining may become unattractive. Such staining can usually be removed, but at some expense. Exposed steel and aluminium will also tarnish with time, according to the chemical composition of the atmosphere resulting from pollution and, hence, the rain that falls on buildings. As with stone, this can enhance the building's appearance or not, in different circumstances. Such colouration of external surfaces and gentle attack from the elements will not usually shorten the life of a façade or adversely affect its performance.

The most common cause of degradation to the performance of cladding systems is the failure of weather seals. The elastic and plastic properties of both profiled sealing strips (e.g. neoprene) and many mastics and silicone sealants change with exposure to sunlight and they can become embrittled. This means the seals are no longer able to accommodate the small relative movements between distinct elements of the façade; they are then likely to crack and become permeable to air and water. After the seals in a cladding system have broken down, they can usually be removed and replaced, giving the façade a considerably extended life. Such work is undertaken by specialist contractors.

The most serious failures of cladding systems occur when the fixing systems fail – due to either the corrosion of steel bolts or brackets, or the failure of load-bearing adhesives bonding fixings to stone or concrete panels or bonding stone veneers to a more massive concrete backing. Single failures can, of course be repaired, but such a failure may be a symptom of a design fault leading to corrosion or poor application of adhesives that may affect many panels in a façade.

Rain-screen claddings made from profiled metal sheet or timber-board cladding are usually not refurbished in situ. Once they have reached the end of their useful life little can be done to prolong it. A long life is best ensured by good detailing that maintains ventilation and prevents the build up of water that causes corrosion or rot.

An existing façade will need to undergo a thorough appraisal of its peformance to assess its condition and refurbishment needs.

Reuse of salvaged or reconditioned products and reclaimed materials

The construction of strong-back and stick cladding systems is such that they can, in principle, be easily dismantled and removed from a building with little damage. Whether this is done or not when a building is demolished or a façade removed depends on the attitude, and perhaps the skill, of the demolition contractor, as well as on the potential value of the products being salvaged. If elements of a cladding system are to be salvaged and reused, storage will also be an issue. Elements designed to be supported in one way when fixed in a façade may not be suited to being stored flat, in piles or stacked resting on their edges. The components or elements most likely to be salvaged and stored ready for reuse are panels of stone such as granite or marble. These are:

- generally in demand;
- likely to degrade little in use;
- easily removed from buildings;
- stored with a low risk of damage:
- possible to fit with bespoke fixings to suit a new cladding design.

Despite some standardization of sizes and fixings, most cladding systems are designed to meet the requirements of a certain building (namely the architect and client), both in terms of column grid and storey height and also visual appearance. It will require some confidence on the part of a salvage company to store façade panels and components to await the right buyer who wants just that size and style of façade. Together these reasons have conspired until now against the reuse of cladding systems.

In practice the reuse of façades is most likely to be successful when it is demand-led. In other words when a design team have decided to reuse an existing façade and search one out early in the design of their building. In this way a practical route to completion can be organized:

- demolition contractors can be properly briefed;
- a storage location can be arranged to cover the period between demolition and installation;
- the rest of the building (e.g. column or wall spacing) can be set out and detailed to suit the cladding;
- repair or refurbishment of components can be scheduled into the cost plan and the construction programme.

Given the popularity of glazed curtain walling in recent years, there is soon likely to be a supply of used glass panels when these buildings are given new façades or demolished. In principle such panels are highly suitable for reuse as they are robust enough to be removed from their original location and stored prior to cleaning and reuse. In practice many such glass units will have a lower performance specification than that required by building regulations when they are to be reused, especially concerning their transmission performance, reflectivity and emissivity.

Nevertheless, if the optical and thermal properties of available glass sheets are known, or can be measured, they might be used in suitable locations in a new building – building regulations now usually prescribe the thermal performance of an entire building envelope, not each square metre of that envelope.

The main issue with reusing glass sheet is whether or not it can be cut to a new size. Ordinary, annealed window glass can be cut to size. Much glass, however, is heat-treated to improve its strength and toughness. Toughened or 'tempered' glass cannot be cut to size. Many building envelopes also use laminated glass. This usually comprises two layers of glass bonded with a polymeric interlayer (usually polyvinyl butyral). If both glass sheets are of annealed (untempered) glass, it is possible to cut sheets of laminated glass, though three cutting operations are needed. If one of the glass sheets is of tempered glass, then the laminated glass cannot be cut.

There also many opportunities for reusing glass panels from a curtain wall in situations where optical and thermal performance is less critical, for example, canopies and in buildings whose internal environment is not that of normal residential or office use, such as airports, railways stations, food markets and exhibitions.

There is already a thriving market in reclaimed profiled metal sheet cladding. This usually takes the form of 'down-cycling' – reuse in a situation that requires lower performance or quality of product. Many agricultural buildings and small warehouses are clad with profiled sheeting removed from buildings. Such panels are identified as having high value by demolition contractors and stored in the knowledge that it will not be difficult to find a purchaser. Metal sheeting has the great advantage that it can be easily cut to shape and machined to suit new fixings or positions of the points of attachment to the supporting structure. The regular, repeating shape of profiled sheet also makes it easy to overlap to any degree to suit column positions.

Timber-board cladding and timber shingles will usually have reached the end of their useful life when removed from a building and it is not practical to refurbish either for reuse. Nevertheless, when the opportunity arises, both timber-board cladding and timber shingles that have been removed in good condition, like all other useful timber, are likely to be reclaimed and available from architectural salvage firms.

Recycled-content building products

Concrete forms the basis of many cladding systems, either as the backing material for a natural stone veneer or as the exposed material itself. While general use of concrete with recycled content is considered in Chapter 3, special mention should be given in this section on façades to the use of 'reconstituted stone'. This form of concrete uses an aggregate incorporating chippings from high-quality natural stones such as granite and marble. When polished, the surface appearance can closely resemble the original stone itself and is highly suitable for use in building façades.

While much post-consumer recycled glass is not clear enough to use in glazed façades that serve as windows, there are many areas of a building façade where slightly tinted glass could be used. Generally, glass used in façades must be toughened or may be a laminated sandwich of ordinary and toughened sheets.

Rigid sheet cladding is available with over 90 per cent recycled content incorporating both post-industrial and post-consumer plastics. Indeed, as the 'plastic lumber' industry grows, it is likely that boards could be made to resemble timber-board cladding.

The US architect Samuel Mockbee, who founded Rural Studio, took a holistic approach to reducing the use of new resources in his buildings. This included the use of rammed-earth walls, reclaimed timber, straw bales and car tyres. He was particularly ingenious in his use of reclaimed materials for parts of the building façade. In a community centre at Mason's Bend, Arkansas, completed in 2000, Mockbee used 80 automobile windscreens salvaged from Chevy Caprices found in a scrapyard. They were purchased at a cost of just US$120 to form an entire façade and part of the roof covering a semi-outdoor part of the building. Mounted on a welded steel frame, they overlap sufficiently to keep wind and rain out, though do not form a fully watertight building envelope. Mockbee used redundant road signs and salvaged aluminium car number plates to create rain-screen cladding in two of his other buildings (Dean and Hursley, 2002).

Roofing

The roofing of buildings is the area of building construction in which reuse and recycling has been most widely taken up. Roofing slates and clay roof tiles in particular are widely reused, not only for roofs but also for wall coverings. This is for several reasons:

- they are simple, primary building materials;
- they are easy to remove during demolition and easy to handle and store;
- they are durable and have a long lifespan;
- they are readily available in large quantities;
- they are available in a variety of types in standard patterns and sizes;
- they have been used for many centuries and can be found in types and styles characteristic of buildings of all ages (to match or suit the rest of an old building);
- they generally require little refurbishment;
- they are often available locally.

One potential difficulty with all tile products is the difficulty of matching new or reclaimed tiles to others in situ. Not only do their colours, size, shape, curvature and thickness vary, so do the many ingenious methods of providing a good interlock between adjacent tiles. The wide variety of tiles means that

Figure 6.1 Reclaimed roof tiles and slates are widely available in architectural salvage yards

Source: Buro Happold

many (especially older ones) may be rather scarce and hence expensive (Figure 6.1).

Other types of roofing and wall-covering materials that may be suitable for reuse or available with recycled content include:

- tiles made of many materials (apart from clay);
- malleable metal sheet (usually copper, zinc, aluminium and lead).

A wide variety of canopies and roofs are made from tensile membranes of woven polyester or Teflon-coated glass fibres. These are always designed and made for unique applications and are not suitable for reuse in a different location. There is an important exception of course – the use of tensile membranes as stand-alone structures for temporary shelter or exhibitions that can be demounted and re-erected in a new location. This type of application is beyond the scope of the book.

The value of reuse

Roofing materials count for a significant proportion of a building's mass and reuse and recycling can lead to great reductions in the demand for new materials (being fired products, ceramic tiles have a large embodied energy).

The principal value of reuse in the case of old buildings is preserving their original appearance. In conservation areas roofing materials are likely to be tightly specified both for restoration and for new build.

When a building is being assessed for its environmental impact using a tool such as BREEAM or EcoHomes in the UK or LEED in the US, credits can be gained by using a proportion (typically 25–100 per cent) of reclaimed/recycled roofing materials (see Appendix B).

Reuse in situ

The majority of tiles and slates are held in position by one or two nails or, for tiles with integral locating nibs, simply their own dead weight. Individual damaged tiles can thus easily be replaced and such roofs maintained for many decades or longer. (Ridge tiles and the end tiles or slates of a roof are often held in place by a cement mortar.)

Malleable sheet roofs and wall coverings are made from copper, zinc, aluminium or lead, and sometimes of stainless steel or even titanium. Each can be repaired in situ should they be damaged by the impact of sharp objects. Metal sheet coverings, especially lead, may also become embrittled with time through regular small movements caused by the wind and eventually crack at a fold in the metal sheet. All metal roofs corrode slowly making corrosion products that depend on the chemical content of the atmosphere and rain. Corrosion is usually very slow, but should it lead to a failure of the sheet, it is likely that the entire roof covering will need replacing.

While tiles, slates and malleable metal sheet roofing and wall covering will eventually need replacing by new roofing, this will be a relatively small cost (and environmental impact) compared to the cost of replacing the structure of the roof beneath. When the roof covering is replaced, reclaimed or recycled materials can be used.

Reuse of salvaged or reconditioned products and reclaimed materials

Using reclaimed tiles or slates can be considered as a viable option for most roofs. To reduce environmental impact (air pollution from transport) reclaimed products should be sourced as close as possible to the building site. While standard tiles are widely available, there is usually a shortage of special tiles such as 'tile and halves' that end each course of tiles, and tiles with air-vents incorporated. New special tiles will probably have to be used.

The quality of tiles and slates can usually be assessed visually. They may have suffered spalling of the edges due to attack from frost. The integrity of a clay tile can be checked by tapping it and listening for a good ring (not a thud).

Many buildings constructed since the early 20th century used roof tiles made of concrete. While many are suitable for reuse, some were made using asbestos fibres (an early form of fibre-reinforced cement) that cannot be reused today. The same applies to the large numbers of garages and sheds roofed with corrugated fibre-cement board that usually contained asbestos fibres. More recent corrugated fibre-cement board is made with non-harmful fibres.

If removed carefully, malleable metal sheet roof and wall covering can be salvaged for reuse. Care needs to be taken to inspect folds for signs of damage and brittle fracture.

Recycled-content building products

Both tiles and shingles with recycled content are available in many materials:

- 'artificial slate' – concrete made with slate aggregate (post-industrial waste);
- resin-bonded crushed slate and other 'waste' materials (however, the resins may have a relatively high embodied energy);
- recycled polymers;
- recycled rubber.

The possibility of specifying metal sheeting with a recycled content (especially alloys of aluminium) should be considered.

Waterproofing

Waterproofing of a building is usually achieved in one of three ways:

1. waterproof mortars, cements and rendering;
2. a continuous flexible membrane;
3. application of a coating in liquid form (e.g. asphalt).

Reuse in situ

Waterproofing products can usually be repaired in situ if the damaged area can be located with certainty.

Reuse of salvaged or reconditioned products and reclaimed materials

The risks of using a reclaimed waterproofing product that is damaged are too high to make it worth considering the reuse of salvaged or reconditioned products and reclaimed materials.

Recycled-content building products

There are many types of roof membranes available that are made from recycled polymers, such as the following typical products:

- a wide range of single-layer polymeric roof membranes are available made from recycled-content polymers – up to 98 per cent;
- damp-proof membranes of recycled polythene (60–95 per cent post-consumer recycled content);
- gas(radon)-proof membranes containing 10 per cent recycled low-density polyethylene;
- preformed drainage layer for contaminated land using 85 per cent post-consumer high-density polyethylene waste.

Asphalt and bitumen are used in many waterproofing systems and can be reclaimed when buildings are demolished. When used again, the recycled material is mixed with a proportion of new material. It is also possible to make roofing felt that is impregnated with a high proportion of recycled asphalt or bitumen. A liquid product, made from rubberized asphalt, used for making waterproof roofing membranes, is available with recycled content. One such material contains around 25 per cent recycled material comprising post-consumer reclaimed rubber and recycled oil.

Case studies: Façade reuse and refurbishment

The Hospital, Endell Street, London

The original building in London's historic Covent Garden area was built as a hospital around 1910. Being located in a conservation area, the retention of the original brick and stone street elevations on the west and south sides was required. It was also decided to keep and reuse the original party wall on the north side adjacent to the end of a terraced row of Georgian buildings and the party wall adjacent to a stained-glass works on the south side of the site. On the east side the visible façade was demolished and replaced, while the basement wall was retained and reused (Figure 6.4).

The two party walls on the north and south sides of the site were retained by steel towers with raking arms built within the site. The Endell Street façade was supported internally by K-braced horizontal flying shores between the west façade and the former stained-glass works to the east. The Shorts Gardens façade on the west side was supported from outside by vertical steel trusses on steel portal frames built over the street. All the retained façades and walls were fixed to their supporting structures by means of ties passing through window openings and clamped internally and externally to the walls (Figures 6.2 and 6.3).

Figure 6.2 The retention structure for Shorts Gardens façade

Source: Price and Myers Consulting Engineers, London

Figure 6.3 The retention structure for the Georgian terrace façade

Source: Price and Myers Consulting Engineers, London

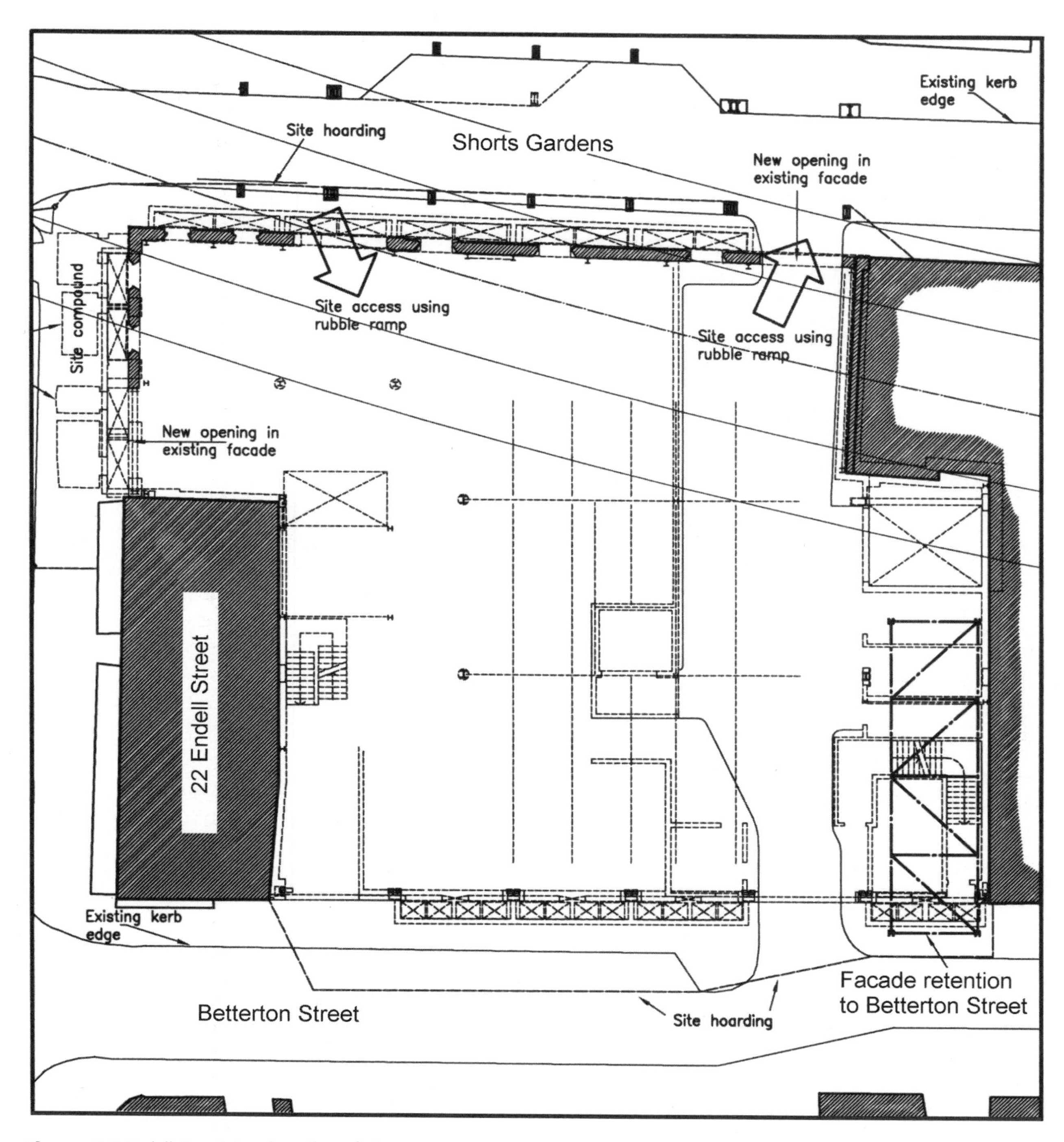

Figure 6.4 Endell Street, London: Plan of site

Source: Price and Myers Consulting Engineers, London

The new development involved digging out a two-storey basement as well as removing the inside of the old building. As the site was bounded on all four sides by fragile masonry façades or whole buildings, it was important to ensure that neither the construction work nor the actions of wind loads or differential heating by the sun led to movements that would cause the masonry to crack. All the temporary works were designed to limit movement to a maximum of 2 millimetres. The movement of the retained façades was monitored using the low-tech, but extremely reliable method of suspended plumb bobs as the datum line from which to measure any movement of the stabilizing towers. During the entire year of construction an acceptable drift of 5 millimetres was recorded, with minor variations probably due to changes in temperature and the stiffness of the soil beneath the structures. Movement of the adjacent buildings was monitored using tell tales fixed across existing cracks in the building fabric to detect any opening of the cracks; no significant movements were detected.

Location: Covent Garden, London
Date: Original building 1910; refurbishment completed 2002
Client: The Hospital Group
Structural engineer for façade retention: Price and Myers Consulting Engineers
Contractor: Griffiths McGee
Refs.: Bussell et al, 2003.

Piccadilly Plaza, Manchester

When Piccadilly Plaza was constructed in 1960 it was the tallest office block in Britain and an early example of a curtain wall façade in Britain using Crittall's Type 'C' Fenestra system. Although this gave the building some architectural interest, its façade was refurbished in situ for wholly commercial reasons. After 40 years the façade was looking rather the worse for wear (Figure 6.5). The milled aluminium mullions were pitted by corrosion and had collected a lot of dirt. The original spandrels, made of wired glass covering insulating panels made with asbestos cement, had at some time been painted to improve their appearance but the paint was peeling badly. Various elements of the façade had been replaced and failed seals had been repaired piecemeal. Years of a low-maintenance regime had left an unattractive façade that reduced the commercial value of the office space compared to adjacent buildings in the prime city-centre location. It was realized that there was little point in refurbishing the building interior unless the façade was also improved.

Figure 6.5 Piccadilly Plaza, Manchester, UK: The curtain wall prior to refurbishment

Source: Buro Happold

An initial assessment of the existing façade proposed a number of alternatives ranging from complete replacement to the minimum possible maintenance. The crucial question was assessing the remaining life of the double-glazing units which, by and large, were still performing well. Balancing the costs and the risk associated with predicting the remaining life, the client opted to keep the double-glazing units and their supporting structure and spend money on upgrading the elements that were detrimental to the appearance of the building.

The visual impact of a number of proposals was studied using computer models. The solution chosen involved overcladding the glass spandrels and replacing the beading that held them in place in such a way that the asbestos panels behind were not disturbed. The original aluminium frame was also overclad with a powder-coated aluminium extrusion made specifically for the purpose and held in place using high-strength double-sided adhesive tape.

The responsibility for the new façade was shared by the façade engineer and the contractor. The contractor warrantied all the new work undertaken while the façade engineers warrantied the existing frame and fixing brackets. Based on the assessment of their condition by the façade engineer, the client has taken on the risk of failure of the original double-glazing units.

Figure 6.6 Several refurbished panels (left) demonstrate the improvement compared to the adjacent existing panels (right)

Source: Buro Happold

Location: Manchester, UK
Date: 1960; façade refurbishment 2005
Client: Bruntwoods
Façade engineers: Buro Happold Consulting Engineers
Architect: Stephenson-Bell Architects
Refs.: Buro Happold

10 Queen Street Place, London

The four-storey concrete-frame office building in Queen Street Place, London was built in the late 1980s with small atria set back from the façade on three sides of the building. The façade consisted of double-glazing units and high-quality granite panels supported on stainless-steel fixings. Plans for redeveloping the building in 2004 involved filling

Figure 6.7 Queen Street Place, London, UK: Granite panels removed from the original façade

Source: Buro Happold

in the floor plates of the original atria and creating new atria, services cores and lift shafts in the centre of the building. In the assessment of the existing façade, a number of granite panels were identified that were damaged or discoloured. Since it was essential that the double-glazing units should be

Figure 6.8 The stainless-steel fixings were also inspected and reconditioned or replaced, as necessary

Source: Buro Happold

Figure 6.9 The new granite façade nearing completion, flanked by the reglazed aluminium framed windows

Source: Buro Happold

warrantied for 25 years a detailed assessment of the existing units, which were already 15 years old, was undertaken. This assessment determined that they could not be warrantied for a further 25 years and so it was decided to replace the glazing throughout the building, installing new double-glazing units into the original aluminium framing that still had 25 years of useful life.

There remained the question of how to deal with the three areas of façade where the atria had been. Rather than scrap the granite panels and reclad the entire building with new stone it was decided to clad the original atria with a new façade of a different appearance from the original and to refurbish the granite panels in the remainder of the original façade.

The granite panels from the original façades to the atria were carefully dismantled, assessed for their potential to be reused and stored on the roof of the building (Figures 6.7 and 6.8). The damaged or discoloured granite panels that had been identified in the survey were removed and replaced by panels that had been salvaged from the redundant areas of the original façade. This strategy led to a considerable saving compared to the alternative of renewing the entire façade (Figure 6.9).

Location: Central London
Date: Original building 1987–89; façade refurbishment 2005
Client: The Blackstone Group
Façade engineers: Buro Happold Consulting Engineers
Architect: Hurley-Robertson & Associates
Contractor: Interior plc
Refs.: Buro Happold

7 Design Guidance: Enclosure, Interiors and External Works

Space enclosure: Partitions, insulation, ceilings, raised floors

The enclosure of space within a building is achieved in different ways for walls, ceilings and floors.

Walls and partitioning

Internal partition walls are formed in one of three ways:

1. Masonry construction of brickwork or concrete block. According to the type of building, a masonry partition wall may be painted, finished with wet plaster or faced with a dry lining of plasterboard which, in turn, may be finished with wet plaster.
2. A lightweight structural frame made from steel or timber supporting a facing made from rigid sheets, usually of plasterboard, plywood or similar products. In older buildings (before the the development of plasterboard around 1900) the facing would have been made using timber lathes and plaster. According to the type of building a partition wall must provide a degree of acoustic insulation between adjacent spaces and provide appropriate resistance to the spread of fire. This is nowadays generally achieved using one or two layers of plasterboard on either side of the supporting frame.
3. Although masonry and partition walls can be removed and rebuilt in a new location to suit changes in internal layout, they cannot be called 'demountable'. To cater for the need to move internal walls frequently, demountable and movable partition walls can be used. Manufactured panels, with suitable acoustic and fire-resistant properties, are fixed between floor and ceiling and may incorporate doors and internal windows. Sliding partitions are also available to allow large rooms to be temporarily subdivided.

In many high-quality old buildings walls were faced with wood panelling that can be removed with relatively little damage during demolition.

Insulation

Thermal insulation is incorporated into external walls as one layer of a multi-layer sandwich. It is usually in one of two forms – a board such as fibreboard, or a loose material that is blown into the cavity, usually between two layers of brick and/or blockwork.

Ceilings and suspended ceilings

The ceilings of domestic and other small buildings are likely to be formed using the same technique as for partition walls – lathe and plaster in older buildings and nowadays, one or two layers of plasterboard faced with plaster.

The ceilings of most recent buildings are formed using ceiling tiles suspended from a metal grid to create a void between the lower face of the floor above and the ceiling. This void usually contains the cables, pipes and ducts required for electric lighting, sprinkler systems and fans and ducts for the supply and extraction of air to/from the room below. An important function of a suspended ceiling is to allow access to the void above for maintenance and fitting new services. The ceiling tiles may be made of many materials including expanded polystyrene, mineral fibre, pressed steel or aluminium. The supporting framework is usually of steel or aluminium, though may incorporate timber too. As with partition walls, the construction of suspended ceilings will reflect the need to provide acoustic and fire separation from the floor above.

Raised access floors

Raised access floors consist of floor tiles usually made from plywood or a hollow steel panel filled with concrete, supported at each corner on an adjustable steel pedestal. They create an easily accessible void beneath the floor to facilitate the distribution of mains electricity, the supply and extraction of air, cables for communications systems, and so forth. As with suspended ceilings, the precise construction of the floor tiles will need to provide appropriate acoustic and fire separation from the floor below. (For timber floors, see Chapter 5.)

Reuse in original building

Unlike the reuse of the load-bearing structure of a building, the continued use of enclosure systems and products falls under the normal processes of maintenance and decoration, which do not fall within the scope of this book.

However, during a major refurbishment of a building, there can often be opportunities for individual components and products to be refurbished and reused rather than stripped out and replaced by new installations. According to the type of product and the ease with which it can be removed, this refurbishment may be done on site or returned to the supplier for reconditioning. It is possible to treat movable partitions and some suspended and raised floor systems in this way. Suppliers need to be consulted for advice on this matter.

Reuse of salvaged or reconditioned products and reclaimed materials

Walls and partitioning

Masonry partition walls can be constructed using bricks or blocks reclaimed from another building. This may be more practical than for exposed masonry in the external wall as partition walls are usually faced or plastered. Hence reclaimed bricks with some visible damage, adhering mortar or remaining paintwork can be used.

Rigid panels of partition walls will most likely be damaged beyond reuse in the stripping-out process during demolition. Timber studwork, however, can be removed with little damage and thus is likely to be available for reuse in a new partition wall.

Wood panelling removed from high-quality old buildings is usually of high value and available from architectural salvage firms.

Suspended ceilings

Enormous quantities of suspended ceilings are removed and consigned to landfill every year as prestigious commercial and other large buildings are refurbished. Both the supporting framework and the ceiling tiles of suspended ceiling systems can be removed with little damage and, in principle, stored, refurbished and made available for reuse. At present, however, there is no significant market for reclaimed, reconditioned suspended ceilings. Designers wishing to reuse suspended ceilings will need to identify a source at an appropriate stage of the procurement process.

Raised access floors

Both the floor tiles and steel pedestals of raised floor systems are usually very robust and easy to remove from buildings without damage. They are highly suitable for reuse, especially in view of the fact that they are supplied in a limited range of standard sizes (typically 600 × 600mm) and are adjustable in height to ensure a flat floor surface.

Recycled-content building products

Masonry partition walls can be constructed of concrete block made with recycled aggregate. Timber studwork can be made using timber salvaged from the demolition of a building.

Gypsum is a naturally occurring material (it is a type of rock) and it is possible to make plasterboard using a man-made gypsum replacement that is a by-product of some industrial processes. Plasterboard can also be made with a proportion of recycled gypsum reclaimed from waste in the manufacturing of plasterboard and waste plasterboard returned from construction sites after damage in transit or as clean off-cuts. Plasterboard is also available in which the facing paper is made of 90–100 per cent recycled paper.

Rigid panels made from recycled timber (e.g. chipboard) can be used for wall panelling, in the construction of fixed partitions, and in the manufacture of movable partitions.

Other examples of RCBPs are:

- Partitions made from a variety of reclaimed and recycled materials including post-consumer and post-industrial timber and post-consumer paper.

- Cellulose fibre insulation made from processed waste paper, usually treated with borax for fire and insect protection. Because of the low-intensity production process it has a much lower embodied energy content than most other insulation products. The product made from post-consumer newsprint is available in two forms – in loose form (around 85 per cent recycled content) that can be blown into wall cavities, and in the form of board (around 25 per cent recycled content).
- Various sheet products with recycled content for partitions between toilet cubicles and around baths or showers. These may range from around 50 per cent post-consumer and post-industrial polymers (usually polyethylene and polypropylene) to 100 per cent post-consumer plastic bottles. Such panel products are also available for making kitchen units.
- Tiles for suspended ceilings made from recycled-content materials.
- Acoustic ceiling tiles made both from post-consumer glass (around 60 per cent) and post-consumer newsprint or cellulose fibre mixed with slag wool recovered from the steel-making industry (20–80 per cent depending on manufacturer).

Windows

Window frames are generally made from four types of material – wood (hard or soft), steel, aluminium, or PVC. When made from timber or steel they are made as a building product that is installed, and can be removed, in one piece. Most aluminium and all PVC window frames are assembled in situ and likely to fall into separate pieces when removed. Many rooflights, screens, louvres and exterior doors and door frames are made using the same techniques and systems as windows.

Wood and metal window frames have long been manufactured in a range of standard sizes, usually based on the size of brick used in external masonry walls. However, since bricks vary in size according to when and where they were made, the corresponding 'standard' window frames will vary in size and may not be interchangeable.

As the aperture to be filled by glazing varies according to the design and cross-section of the frame, it is cut to size. With the increased use of double glazing, some standardization of sizes has occurred, but not to a great extent. Sealed double-glazing units of various thicknesses can be made to suit any size of window aperture; however, many wood and metal window frames are unable to accommodate the increased thickness of sealed double-glazing units.

Most metal and some wooden window frames incorporate weather seals made of polymer extruded to fit precisely in grooves in the frames. These seals can suffer damage and polymers may deteriorate with time after exposure to sunlight.

The growing concern for reducing energy consumption has lead to an increase in the use of blinds in recent years. External blinds or blinds within the cavity of double-glazing units are most effective because they prevent solar energy from entering and overheating a room in summer. On the other hand, internal blinds are easier and cheaper to fit. Blinds are also useful for reducing glare when the sun is low in the sky.

Internal and external shutters were (and are) often used to keep the sun out as well as providing added security to a building. In domestic buildings they are generally made from timber and tend to be used in Mediterranean and older Georgian buildings.

Perhaps the most significant feature of windows is their function as a barrier to both noise and heat (or cold). The thermal and, to a lesser extent, acoustic performance of windows has been rising steadily since the mid-20th century, driven largely by regulation and legislation to reduce energy demand and improve the level of comfort in buildings. In many countries, building owners are required to upgrade the thermal performance of windows when they are replaced or even when they are reglazed.

General issues regarding reuse of windows

- Ensure there is an adequate supply of the correct type, material, number and size of windows before the project reaches detailed design.

- Ensure the windows meet current standards and legislative requirements, or upgrade to suit. This may affect both the frame and the glazing, which may need to be tinted to reduce glare or treated with a low-emissivity coating.
- It may be a requirement that individual windows or the entire building façade are tested after installation or refurbishment for air-tightness, water penetration or acoustic performance.
- Consider the whole life of the construction elements, including potential energy savings, the residual life of the window and refurbishment costs. In some cases, this may result in recycled-content or new materials being more applicable to the project than a reclaimed window.
- Durable hardwood timber frames are a good environmental option as they generally require less maintenance and repair than less durable softwood frames.
- Where the timber frames are not entirely reclaimed or made using recycled-content materials, specify other timber to originate from an accredited sustainable source (with FSC or equivalent accreditation).
- Suitable catches and locks and other architectural ironmongery can be obtained from architectural savage firms.

Reuse in the original building

Value of preserving original windows

Windows are often described as 'the eyes of a building', not only because the occupants look out, but because the gaze of outsiders naturally falls on windows in the way that we instinctively look at the eyes of people we meet. The character of a building is, then, strongly influenced by the type of window used and its original character is preserved by retaining the original windows for as long as possible. This has a direct (positive) influence on the value of the property.

Prolonged use in situ

Most types of window can be maintained and refurbished in situ over many decades. Softwood and hardwood frames require regular repainting and may require rotten sections to be replaced. Unpainted hardwood frames require regular treatment as protection against the elements. Old steel-framed windows require painting (less regularly than timber) and corrosion may need to be treated. Newer steel-frame windows are powder-coated and can perform for several decades without attention. Aluminium window frames corrode slowly and lose their metallic lustre quickly. Refurbishment in situ is usually unsatisfactory.

The opening and locking mechanisms of all windows are vulnerable to damage but they can often be repaired in situ, or replacement parts can be obtained to prolong the life of the window.

When windows are refurbished in their original location, many window frames made to carry single glazing (4–6mm thick) will be unable to accept sealed double-glazing units (12–25mm thick). This will make it difficult to comply with more stringent building regulations that demand improved thermal performance of windows.

Metal-framed windows in need of refurbishment can be removed, reconditioned by specialists and refitted in their original location.

Warranty

Suppliers of new (replacement) sealed double-glazing units usually give warranty against failure of the seal.

Reuse of salvaged or reconditioned products and reclaimed materials

Value of reusing windows

In an old building the principal value in using reclaimed, reconditioned windows is to preserve its original character. Sadly, many buildings have been ruined by the installation of windows of inappropriate design.

Otherwise, the use of reclaimed windows is likely to be selected because it is cheaper or in order to incorporate recycled-content products in the building.

Availability

Complete windows are widely available from architectural salvage firms. Apart from the condition of the frame, handles, stays and locks, it is important to assess the condition of seals and gaskets and, if damaged, establish whether they can be replaced.

The main considerations for reclaimed shutters and blinds are the size and design of the window. If possible windows should be obtained with their matching shutters or blinds, which should also be inspected to ensure they are in working order.

Recycled-content building products

Availability

Window frames are available made from recycled timber and polymers.

Recycled (virtually remanufactured) sealed double-glazing units are available. Salvaged units in which the seal has failed can be dismantled to liberate the two sheets of glass. These can be cleaned, cut to size if necessary, and used to make a new sealed unit.

Recycled (re-melted) glass is not currently specified for window panes as it tends to suffer from slight coloration due to impurities.

Doors

Like windows, doors and their associated fittings contribute greatly to the character of a building and many owners of old houses go to great trouble to preserve or reinstate original doors and door frames, or faithful replicas.

Domestic external doors are invariably of timber, often with an appearance that is characteristic of their date of manufacture – Georgian doors, for example, usually have six panels while Victorian doors have just four.

Modern internal domestic doors are often made as a sandwich of a lightweight honeycomb made of cardboard faced by sheets of plywood or hardboard.

Doors in non-domestic buildings, including fire doors, are usually more substantial than domestic doors. Many are made of wood, though glass doors are increasingly popular when forming part of a fully glazed building façade.

General issues regarding reuse of doors

Most of the issues relating to the reuse of doors are the same as for windows:

- Ensure there is an adequate supply of the correct type, material, number and size of doors before the project reaches detailed design.
- Doors are nowadays made to standard sizes though may have been trimmed to fit a particular door frame.
- For non-domestic buildings doors must now be wide enough to comply with the requirements of access by disabled users of buildings, that is, sufficient width and an absence of steps. Purchasers of domestic doors may also wish to address these considerations.
- If possible, obtain the door frames made for the doors being used.
- Suitable locks and other architectural ironmongery can be obtained from architectural savage firms.

Reuse in the original building

Doors can be repaired and refurbished in similar ways to windows.

Reuse of salvaged or reconditioned products and reclaimed materials

Availability

Domestic doors are widely available in architectural salvage yards.

At present there is no market for doors reclaimed from non-domestic buildings being demolished or refurbished, though in fact many currently scrapped would be highly suitable for reuse. This is partly the inertia of the industry and partly because of the need to be confident of a door's fire rating in order that a building may obtain insurance cover.

Timber doors may require rehinging, conditioning (stripping or dipping), removal of nails and such like, treating, sanding, painting and varnishing (Figure 7.1).

Recycled-content building products

Availability

Doors are available made from recycled timber, paper, plastic (some are a combination of different materials). Moulded hardboard doors have become a common choice for interior doors and are a good use of waste shavings from timber-mills. The design team should select a material with a high recycled content to ensure the greatest environmental benefit.

Recycled-content doors and door frames are available made from timber composites and recycled polymers. For example, many doors are made using timber panel products, such as MDF, that is generally made using post-industrial wood waste such as sawdust. Also, many modern lightweight doors consist of a honeycomb, made from paper or card, sandwiched between two panels of hardboard. The paper/card honeycomb can be made from 100 per cent post-consumer paper fibre and the hardboard itself uses post-industrial timber waste.

Figure 7.1 Revolving door

Source: Big Old Doors (www.bigolddoors.co.uk)

Figure 7.2 Stone staircase (*c.*1700) at Hampton Court Palace, near London

Source: Price & Myers Consulting Engineers

Stairs and balustrades

Both old and modern domestic buildings generally have timber staircases that are built in situ and can only be removed from their location by complete dismantling.

Many old masonry buildings have stone staircases that are built into the fabric of the building. There are two types. In one the tread, whether straight or the characteristic key hole shape of a spiral staircase, spans between supports at both ends. In the other type, called cantilever staircases, the tread appears to be supported at only one end, seeming to cantilever out from the wall. In fact they do not work as cantilevers but are supported by the tread beneath and prevented from twisting by being built into the wall to a depth of just 100–150mm. These remarkable staircases were first recorded in Palladio's *Four Books on Architecture* published in 1570, and introduced into Britain by the architect Inigo Jones in around 1613 (Figure 7.2).

Many buildings, old and new, have external fire escapes made of iron or steel.

Reuse in original building

Generally the condition and safety of all types of staircase can be assessed in a survey of the building fabric, and remedial action taken if found to be necessary, for example due to excessive wear, fracture of treads, rot or infestation.

Sadly many fine stone cantilever staircases have been removed from buildings, having been deemed to be unsafe by modern structural engineers who did not understand how they worked. Fortunately clear guidance has now been published explaining their action and how their safety can be demonstrated as well as guidance on how they can be repaired, if necessary (Price and Rogers, 2005).

Reuse of salvaged or reconditioned products and reclaimed materials

It is unlikely that entire staircases can be reused as there is such a variety of staircases in the type, length of stair, floor-to-ceiling height and corridor width.

Although entire timber staircases are seldom available, their decorative joinery, banisters and newel posts are often to be found in architectural salvage yards.

It is in the nature of most cast-iron spiral staircases that they are fitted into buildings after their completion, being tied into the main fabric at a minimum number of places. For this reason they can easily be removed and are not uncommon in architectural salvage yards, usually as a heap of pieces that require careful reassembly. Modern spiral staircases of steel can also sometimes be found and, if the design team is aware that a freestanding steel staircase is available, then the areas of circulation could be designed to incorporate it.

A further source of staircases, sometimes in cast iron, are the external fire escapes that were often fitted to buildings in the 19th century when fire safety regulations demanded their addition to multi-storey buildings that previously had only one means of escape (Figure 7.3).

A great many balconies and balustrades made of cast or wrought iron have survived from Victorian

Figure 7.3 Cast-iron staircase immediately prior to reclamation and reuse

Source: Big Old Doors

times and these too can occasionally be found in architectural salvage yards. Depending on their size and the nature of the loads they need to carry, a structural engineer may be required to assess their capacity and suitability.

Recycled-content building products

The main opportunities for constructing staircases with recycled-content materials is using concrete made with recycled aggregate or timber products such as chip and blockboard.

As with many parts of a building, there are no doubt occasional opportunities for an imaginative use of reclaimed materials to construct stair banisters and railings to balconies – old bedheads perhaps, or other pieces of metal found at the scrapyard.

Surface finishes/floor coverings

It is in the nature of floor coverings that they are worn with use, either by footfall or furniture as well as the action of repeated cleaning. In general, coverings such as carpets that wear quickly can be removed and replaced easily. Those that are more durable, such as screeds and woodblock floors, are less easy to remove; indeed, screeds should be considered as permanent.

Reuse in situ

It is only hard wearing floor coverings that are able to be refurbished to a useful degree in situ:

- Some screeds can be filled and re-polished.
- Stone flag floors can be cleaned and polished and individual damaged flags can be replaced.
- Tiled floors too can be repaired in situ, so long as suitable replacement tiles can be made to match the colours or patterns of a decorative feature. Individual damaged quarry tiles can usually be lifted and replaced.
- If they are thick enough, woodblock and other timber floor coverings can be sanded, polished and refinished with varnish or wax. If there is significant damage they can also be easily lifted and re-laid.

Reuse of salvaged or reconditioned products and reclaimed materials

Apart from screed floors that are applied as liquids, most floor coverings can be removed with relatively little damage allowing them to be reused elsewhere. Tiles fixed with modern cements or chemical adhesives, however, are more likely to be damaged in their removal. Whether a floor covering is reclaimed depends on the condition of the covering and the quality of their decorative features – a tiled mosaic or oak block floor is likely to be in greater demand than a second-hand office carpet.

Salvage yards and other dealers in second-hand building materials are likely to stock a wide range of reclaimed floor coverings:

- stone slabs and flags;
- bricks, setts and cobbles;
- ceramic quarry tiles;
- tile mosaics of ceramics, granite or marble;
- woodblock, parquet and other timber floors;
- carpet, carpet tiles;
- underlay and acoustic-isolation rubber matting.

Flags, tiles, etc.

Traditional stone flags are easily lifted and prized for their natural appearance and durable qualities. They can be readily found in salvage yards. If their previous location was in an industrial building, which it often was, it is possible they will be stained by oils from machinery. Floors made of brick or stone setts often have a bedding of cement that may make them more difficult to lift, though the pieces themselves are not likely to be damaged. Bricks and setts laid before the early part of the 20th century were bedded in lime mortar that does not adhere strongly and so can more easily be cleaned off prior to reuse.

Tiles made from ceramics, terrazzo, granite and marble, laid on a mortar bed, provide particularly durable floor finishes and supplies of reused stock are often available. As all these materials are brittle and often of low strength, great care needs to be taken when they are lifted and transported.

Timber floors

A reused wooden or parquet-type flooring, treated with natural oils or waxes, is a good environmental option. Woodblock, strip and mosaic floors tend to use softwoods, whereas parquet flooring is often made from tropical hardwoods. The hardwoods commonly available from reclaimed sources include oak, maple and teak, and they provide a durable hardwearing surface with a substantially longer life expectancy than softwoods.

Many reclaimed timber boards are suitable for laying as timber flooring; however they will usually need to be checked for infestation and rot and have nails removed. Just as with virgin timber, reclaimed timber will require cutting to length, sanding and finishing.

Carpet and carpet tiles

Both carpet and carpet tiles can be easily removed from a building and reused. A healthy second-hand market flourishes and the reuse option can be especially attractive when relatively small areas are involved. The condition and potential remaining life of carpets and carpet tiles are easy to assess. Some carpet manufacturers offer a reconditioning service, though this would probably be viable only for high-quality carpets. Underlay is usually more difficult to remove from a building without damage as it often adheres to the floor or lining paper on which it was laid. Being of softer material it also often wears more rapidly than the wearing surface laid on top.

Flexible sheet materials

A variety of synthetic and natural sheet materials are used for floor coverings. Their durability varies considerably and they are often glued to the substrate on which they are laid, making their removal without damage virtually impossible. If they are not glued down and are sufficiently durable they can, like carpets, be easily assessed for their potential reuse. Many of the synthetic floor coverings contain polyvinyl chloride (PVC or vinyl), which some people regard as an unacceptable choice on environmental grounds.

Linoleum is worth special mention. It is made by oxidizing linseed oil to form a thick mixture called linoleum cement that is mixed with pine resin and wood flour and bonded to a jute backing to form a continuous sheet. Not only is linoleum a flooring material made from natural, renewable ingredients, it is also very durable and has a long life. As long as it can be lifted easily, and is not damaged during handling and transportation, it can be cleaned and is highly suitable for reuse.

Recycled-content building products

A wide variety of floor coverings with recycled materials content is available:

- Carpets with a recycled content of post-industrial and post-consumer waste. (It should be noted that some foam backings for carpets are blown with environmentally damaging chlorofluorocarbons (CFCs) or hydrochlorofluorocarbons (HCFCs).)
- Some synthetic carpets use a backing material made from recycled polymers, typically 70 per cent by volume, 50 per cent by weight.
- Carpet tiles with around 20 per cent post-industrial scrap fibre in the wearing surface of the carpet.
- Carpet underlay made from post-industrial scrap and post-consumer rubber tyres with a 100 per cent recycled content.
- Felt underlay made from rags, wastepaper and/or wool fibres.
- Rubber floor tiles made from post-consumer rubber (usually car tyres) with a recycled content of 60–100 per cent.
- Polymer floor covering in sheet form, made from post-consumer polythene terephthalate (PET).

Wallpaper and paint

Although there is not yet a wide range of options, it is possible to obtain both wallpaper and paint with recycled content. Backing or lining paper is available made with 100 per cent recycled paper. Latex paint with a recycled content of 50–90 per cent post-consumer material is available in a range of colours for both indoor and outdoor use.

Furniture and equipment

Choosing the reused or recycled option for furniture, fixtures and fittings and installed equipment is a particularly sensitive issue since the users of buildings come into close contact with them. This may have both negative and positive consequences. Among the disadvantages are that furniture and fixtures are usually chosen for aesthetic reasons as well as functional. Reclaimed components may be slightly imperfect or irregular and it may not be possible to meet the needs of a whole building with items of a consistent or homogeneous style. A conscious decision to specify reclaimed and recycled materials will also reduce the range and choice available.

Alternatively, the users of a building may wish to use such furniture and fittings as a means of conveying some of their organization's values. The visual impact of the finished building will raise awareness of the recycled and reused components and can serve as a marketing tool, particularly where the building is used by the younger population who may take such expectations into their future working and home lives. Many organizations now have environmental policies that include a commitment to reuse and recycle goods and materials whenever possible. Reclaimed furniture and fittings can provide highly visible evidence of a firm's environmental commitment.

The range of second-hand goods available in this category of products for buildings is large, and the examples below represent only a selection to illustrate the variety.

Reuse in situ

Apart from prolonging the life of goods by normal cleaning, repair and maintenance, there is little to say concerning the reuse of furniture and equipment in buildings. One exception to this general rule is the possibility of re-enamelling baths in situ.

Generally it will be easy to assess the condition of the goods, though catering and other equipment will need to be tested to ensure it functions satisfactorily.

Reuse of salvaged or reconditioned products and reclaimed materials

There are likely to be good opportunities to find all types of furniture and equipment for reuse both in buildings due for demolition and from the many dealers in second-hand and reconditioned furniture, equipment and other goods. Trade journals have long made finding such goods relatively easy, and the internet has made it even easier. Furniture and fittings of historic interest are in great demand.

Furniture

In all countries there are many dealers in second-hand furniture for both domestic and small-scale commercial use. When larger quantities are needed, it can be a viable option to find goods through the auctions that are held when large buildings are being emptied prior to the soft strip and subsequent refurbishment or demolition.

Equipment

A great deal of catering and cleaning equipment is available on the second-hand market and through building-clearance auctions and many countries have dealers who trade exclusively in 'government surplus' goods and equipment. This can cover every imaginable item that may be needed to equip or furnish a building (and a lot more artefacts besides).

As with the reuse of all equipment, it will be important to try to establish the condition, performance and safety of items prior to purchase in order to ensure compliance with current regulations and legislation. Cleanliness will be an especially important issue for all goods for use in kitchens and bathrooms.

Recycled-content building products

Recycled-content building products in the furniture and fittings area are likely to be made of two types of material – recycled polymers and materials containing cellulose fibres (wood or paper).

The former group includes items such as the following:

- moulded furniture made from recycled polymers;

- furnishings, made from recycled materials, which also provide protection from everyday wear and tear;
- various panel products with recycled content are available for making kitchen units. These range from around 50 per cent post-industrial and post-consumer polymers (usually polyethylene and polypropylene) up to 100 per cent post-consumer plastic bottles;
- a variety of indoor and outdoor signs can be made from recycled content polymers. The percentage of recycled content varies considerably with the manufacturer.

Products made with recycled-content materials based on cellulose fibres include:

- Kitchen units of a material such as Tectan, made from compressed paper drinks cartons (see Figure 2.12).
- Furniture made from various timber-based panel products such as chipboard, hardboard, blockboard, medium-density fibreboard (note that some of these materials have a high environmental impact because of the polymer matrix or adhesive used to bind the timber particles or pieces).
- One US manufacturer currently offers a material with recycled content that can be worked like timber and used to make fireplace surrounds (Quinstone). It contains 40–50 per cent post-industrial paper fibre mixed with resin and gypsum. Various mouldings, panels and other profiles are available.

Other materials available include a variety of recycled materials that can be used for kitchen work surfaces. These may be in the form of tiles made from recycled ceramics and glass, or as a material with the characteristics of lightweight concrete. The latter can be made from recovered fly ash (up to 20 per cent) as a cement replacement or with 30–40 per cent post-consumer materials (for example polymers) as the filler material. It can be moulded in situ or made into cast products such as sinks and shower bases. One example, available in the US, is Syndecrete.

Sanitary, laundry and cleaning equipment

Sanitary, laundry, cleaning equipment includes:

- sanitary equipment (washbasins and toilets);
- other fittings linked to water supply and removal systems;
- laundry and cleaning equipment.

When buildings are refurbished or demolished, sinks, baths and sanitary appliances, and their various fittings are regularly removed for reclamation, especially when they are old or characteristic of a particular architectural era such as Victorian, Edwardian or art deco. In addition to normal cleaning operations, enamelled baths can be refurbished by re-enamelling. Specialist salvage dealers clean and refurbish them before offering them for resale. At present few modern items are salvaged because they are not in demand and the cost of cleaning and refurbishment is high compared to the cost of buying new goods. This situation is likely to change as demand for goods for reuse increases.

Many sanitary, laundry and cleaning components are not reused or recycled as it is not economically viable due to difficulties of reclaiming the equipment without damaging it. Removal is more costly than obtaining new equipment. That said, sanitary equipment generally comes in standard sizes and so allows reuse.

There is a small industry for reclaiming toilet pans and washbasins for reuse. A good architectural ironmonger should know of local yards specializing in recycling sanitary products such as sinks and toilet bowls. These are generally reclaimed from demolition sites or removed during a refurbishment.

For larger buildings there may be problems getting the quantity of one particular design of toilet pan, cistern and washbasin, but as long as the quantity is relatively small it should be feasible and acceptable to reuse toilets and washbasins.

All reused sanitary equipment would need to be in compliance with the water regulations. Toilets, for example, need to meet the prevailing maximum flush volume of six litres. Specifiers should also check the required flush volume before reusing toilet pans in a new building, as older bowls may not operate effectively with smaller flush volumes. A dual-flush mechanism may also be appropriate.

Proprietary seals for older sinks and toilet bowls may no longer be available, and the modern equivalent may be of the wrong shape and/or diameter. Even if they appear to fit, new seals may fail after a short period due to stress or because they are a slightly looser fit. Contractors should be required to carefully assess such issues and be asked to propose a (warranted) solution.

Quality will be an issue. As sanitary products are highly visible items, damage should be minimal or non-existent. Client philosophy and acceptance will determine the quality expectation. Various ancillaries may need to be replaced, such as taps and flushing mechanisms, often for purely aesthetic reasons

Whether for reuse in situ or in a different location, mixing and matching existing sinks and pipework with new taps (and combinations thereof) will require careful attention to services coordination.

Systems containing a declared recycled content are not generally available for sanitary ware and cleaning equipment.

External works

The external works around buildings include many products and types of construction making use of a wide range of materials. Being external and often open to public access, both durability and security are important issues. External street furniture must be both vandal-proof and secure.

Reuse in situ

Apart from prolonging the life of external works by normal repair and maintenance, there is little to say concerning their continued use or reuse in situ.

Reuse of salvaged or reconditioned products and reclaimed materials

Fill and topsoil

Probably the greatest opportunity for reusing materials during external works is the use of crushed masonry generated during demolition as fill or for landscaping purposes. Most convenient of all is when this material has come from the demolition of a building on the site itself, since this avoids transport costs and the environmental impact of transport. Such material may have to be graded to suit its purpose and it must be free from contaminants that might otherwise lead to pollution of ground water as a result of leeching. The reuse of such material may be awarded credits in an environmental assessment tool such as BREEAM, LEED or EcoHomes (see Appendix B).

Likewise, topsoil excavated from a site can be used/reused in landscaping on the same site or another, as long as it can be conveniently stored until needed. This can lead to a significant saving in cost and transport, as well as a reduction of environmental impact. The soil itself can be improved by the use of wood shavings or other organic waste or composted biosolids.

It is important to note that even where previously developed land is contaminated it is still possible to 'recover' the contaminated material in situ using one or more remediation techniques. The UK Building Regulations 2000 (2004 Edition) Approved Document C (*Site Preparation and Resistance to Contaminant and Moisture*) summarizes overall requirements. In addition the UK Environment Agency Contaminated Land Research Report CLR 11 (*Handbook of Model Procedures for the Management of Contaminated Land*) gives useful guidance on the selection of appropriate procedures and remediation techniques.

Paths and roads

There are many methods, both proprietary and generic, of constructing paths, roads and car-parking areas using slabs, setts, cobbles and interlocking paving, bricks and blocks. Most of these are easy to lift and relay in a new configuration on the same site or reuse on another site.

Outdoor/street furniture and equipment

External furniture is required to be secure, weather-resistant and durable. The external landscape of a building is often open for public usage, and thus the materials are required to be more durable than standard installations. Some recycled cast-iron external furniture is available with graffiti-proof paint for low maintenance. Security is also an issue so the fittings and furniture cannot be stolen and transported off site. External furniture is often bolted down to the foundations below the ground or stainless-steel security fastenings are used.

Reusing a component is suitable dependent on the condition of the item. Most components will require visual, structural and safety testing to determine the work required prior to installation. A comprehensive specification is required to ensure the appropriate work is carried out and the associated cost implications. The component will be required to be weatherproof, secure and low maintenance.

Recycled-content building products

There are many opportunities for using products made from two types of recycled material – concrete made with recycled aggregate and recycled plastics.

Recycled-content concrete products, made with recycled aggregate, include:

- paving stones and blocks;
- kerbs and edgings;
- ornamental features;
- garden and street furniture including benches, barriers and bollards.

Large pieces of reclaimed timber, such as old railway sleepers and telegraph poles, can be a useful raw material for making various external goods such as car-park barriers, benches and playground equipment (see Figure 7.5).

Figure 7.4 Rubber recycled from car tyres being laid to form an artificial surface for outdoor sports

Source: Bovis Lend Lease

Paths and roads

The paths, roads and car-parking areas around buildings offer opportunities for using recycled materials. Their sub-base can be made using crushed masonry recycled from demolition. The wearing surfaces can be made using in situ concrete made with recycled aggregate, concrete blocks made with recycled aggregate, or recycled asphalt incorporating granular fill of crushed recycled masonry or crushed glass. Asphalt paving is available containing a small proportion of recycled asphalt (10–15 per cent), which will reduce demand for virgin aggregate and for the virgin petroleum used to make asphalt. Recycled rubber from car tyres can also be incorporated as a binder so increasing the recycled content to 30 per cent or more.

Paths with lighter loading can be finished with a layer of gravel made from crushed recycled masonry or woodchip made from recycled timber.

Special surfacing

The surfaces of children's playgrounds and artificial playing surfaces for various sports can be made from products incorporating recycled rubber, usually recycled from car tyres (see Figure 7.4).

Outdoor/street furniture and equipment

A great many recycled-content products for external use are available made with recycled polymers. In the US these materials are often called 'plastic lumber' and are available in a range of colours. Examples include fencing and barriers, benches, picnic tables and other park furniture, playground equipment, playground and sports playing surfaces, decking and flooring and garden furniture. Some typical specific examples are:

- Synthetic surfaces for made from used tyres (85–95 per cent post-consumer rubber) are available for outdoor sports-playing areas for tennis, hockey, football and so on, as well as for children's playgrounds. Various degrees of cushioning and drainage are available. Synthetic surfacing made from recycled PVC (e.g. insulation from electric cables) is also available.
- A wide variety of garden furniture, fences and gates, decking and decorative features used in garden layouts can be made from 90–100 per cent recycled polymers. This plastic lumber or plastic wood is very durable, available in many colours and requires no maintenance. Manufacturers often stress the life-cycle cost-benefits of plastic wooden products over natural wood or metal alternatives. Its strength and stiffness varies according to the type of polymer used. The density and strength can be increased by including up to 50 per cent post-industrial wood fibre with the recycled polymer. An even stronger material is obtained by mixing up to 20 per cent post-industrial glass fibre with the recycled polymer to create a fibre-reinforced plastic.
- Recycled-content polymer products, made from old carpets for example, can be used for various applications in car parks, such as wheel stops, traffic and pedestrian barriers and signage. They have the advantage over concrete that cars are not damaged in minor collisions.
- Sound walls or acoustic fences serve to absorb the sound of vehicles or trains and so to reduce the disturbance to adjacent properties. These can be made using shredded rubber from car tyres sandwiched between outer walls of timber or concrete. A 4-metre-high sound wall might use 70 tonnes of recycled rubber per kilometre length of wall.

Figure 7.5 Reclaimed timber in Crumbles Castle, a children's adventure playground near Kings Cross, London

Source: Bovis Lend Lease

Case Study: Plastic lumber

The 1999 reMODEL project was carried out to showcase new technologies developed for the construction industry. Completed for the 1999 National Association of Home Builders Remodelers' Show, the project involved the complete refurbishment of a brick masonry row house built in the 1930s in downtown Philadelphia.

The owner and builder had extensive previous experience in renovating similar buildings throughout Philadelphia, and was asked to incorporate seven new technologies into this refurbishment for evaluation. One technology assessed was the use of a wood-polymer composite lumber called Trex®, chosen as the decking surface for the rear stoop or porch of the house.

Composite lumber products are made from a combination of recycled plastic materials (e.g. plastic bags and reclaimed pallet wrap) and waste wood. The plastic component of the material defends against insect and moisture damage, while the timber component provides a natural feel while protecting against ultraviolet damage. These products are generally only used as a replacement for timber in non-primary load bearing applications, such as decking and for garden furniture, and come in a range of standard sizes and sections. For the rear stoop the contractor used 2 × 6 inch sections of composite lumber that were cut to size and fastened on site. Cutting was carried out with conventional tools and galvanized woodscrews were chosen to fasten the boards, although almost any conventional fastener will work with this product. After the boards were fastened, the deck was complete and ready for sanding, painting or staining. These treatments are not essential for wood composite products as they do not require sealants for protection, and would only be carried out for aesthetic purposes.

The installation of the deck, including work carried out to reattach the underlying frame, required the services of one carpenter and one labourer for eight hours. The total labour rate charged was US$45 per hour, and with the cost of Trex® lumber at US$20.89 per section, the total cost for the job was approximately US$485.34. The same deck built with conventional lumber would have cost approximately US$76 (or 16 per cent) less as a result of lower material costs. While recycled wood/plastic composite decking is more expensive than the conventional pressure-treated wood, the lower maintenance and increased durability mean this product can be a viable alternative in many outdoor situations.

Location: Philadelphia, PA, US
Date: 1999
Client: Mark Wade
Contractor/remodeller: John Fries
Sponsorship: National Association of Home-Builders (NAHB), and Partnership for Advanced Technology in Housing (PATH), US
Ref.: www.ToolBase.org/docs/ToolBaseTop/FieldResults/2954_ModelreModel.pdf
Website: www.ToolBase.org; www.pathnet.org

8 Design Guidance: Mechanical and Electrical Services

Mechanical and electrical services

Reuse, reclamation and recycling of mechanical and electrical services

Both in their design, and in relation to reuse, reconditioning, reclamation and recycling, mechanical and electrical services have much in common. They also share much that sets them both apart from other building elements dealt with in previous chapters. The key difference is that building services comprise two fundamentally different types of goods. There are many fixed items such as pipework, ducts, cabling and the various racks and frames to which the means of distribution are fitted, which can be treated in much the same way as other products fitted in buildings such as windows, partitions, sanitary ware and so on. In addition is a wide variety of working machinery, plant and equipment that, *as a matter of course*, is subject to a regime of regular repair, maintenance, replacement and recommissioning. Most building services systems and equipment are regularly tested to ensure that they meet a defined set of performance targets. Large pieces of equipment are designed to be repaired and maintained on site. Smaller items that fail to perform adequately will be removed and replaced as a matter of course. Building services equipment is also, therefore, usually easy to remove from a building.

Many components, such as generators, motors, large electrical cables, fan-coil units and transformers, are eminently suitable for reuse once they have been refurbished. There are also items that are not subject to wear, such as cable trays, that can be easily dismantled, cleaned and reused. Unlike many materials and goods that are removed from buildings, such as brick and concrete, building services equipment is manufactured from high-value metals and plastics that can be put back into the supply chain for recycling. The scope for recovery and reuse of building services is therefore very high given the right infrastructure, and commercial and political will. In many cases, all that is needed is enough motivation.

Another consequence of these characteristics of building services is that the guidance presented here on specifying reused goods or recycled-content products in relation to new construction can be equally useful for those engaged in building refurbishment and those who buy machinery, plant and equipment as part of the normal programme of repair and maintenance for a building.

Considering that some 30 per cent of the total cost of a building will involve building services, it is reasonable to assume that a similar proportion of construction waste is in the form of old boilers, air-conditioning equipment, electrical cables and dozens of other mechanical and electrical items.

Legislative tools to encourage reuse and recycling already exist in many countries. In Europe, the Waste Electrical and Electronic Equipment regulations, and the Restriction of the Use of Certain Hazardous Substances directives have recently been introduced to enforce waste-management policies and to foster a positive attitude to reuse, reclamation and recycling. Regulations such as these are a good start, but there is still a long way to go. The public is also increasingly behind moves to reduce unnecessary waste, and corporate responsibility heads the list of issues that matter to many consumers when they are forming an impression of a company. Nevertheless there are many hurdles to overcome and this will require changes of attitudes among the many players involved:

- Building services products are often considered to be waste rather than reusable or reclaimable.
- Refurbishment of equipment is often not practicable because product lives are long compared to the rapid development of technology.

- The marketing of refurbished products is difficult as they may be (wrongly) considered as second-hand and therefore have a shorter life.
- Extended supply chains and sales through trade outlets and online distributors make it difficult to track products through their life to ensure manufacturers can reclaim products for reuse or reconditioning.
- Recovery of refrigerants and oils (under existing legislation) needs to be effectively policed.
- The inconsistent purity and quality of recycled materials, such as plastics, can make it difficult to meet specification requirements and national regulations.
- The large number of materials available for reuse and recycling needs to be rationalized (reduced).
- There is poor knowledge of the availability and pricing of reclaimed building-services equipment.
- Legislation and regulations may affect the potential for product reuse, such as the need for boilers to meet requirements for increased energy efficiency.
- It can be difficult to keep up with the vast amount of relevant legislation and regulation, which is often difficult to interpret and arrives from many different sources.

The majority of building-services plant and equipment is specified by building designers who rely on products and information provided by manufacturers of building services equipment. The designers cannot usually influence how the products are made. There is thus a great deal that can be done by manufacturers of building-services equipment to increase the amount of reuse, reconditioning and recycling that goes on in the industry, and there is great commercial potential for many of these actions (BSRIA, 2003). As with many other manufactured products, such as cars, there is room in the market for manufacturers who promote their responsibility to the environment (Box 8.1).

Box 8.1 Key actions for manufacturers of building-services equipment to facilitate reuse, reconditioning and recycling

- Provide a low-cost product return service.
- Stamp materials and product mouldings with product and materials information using bar codes, labels, ink stamps and laser-jet etching.
- Make use of 'eco-labels' displaying recycled-material content and potential for reconditioning as positive features to encourage product reuse.
- Promote plant-leasing arrangements whereby manufacturers retain ownership and responsibility for products. These improve the opportunities for reconditioning and refurbishment, since the manufacturer is more likely to be willing to carry the risk and financial benefit of product reuse.
- Where the manufacturer maintains the installation, it is easier to return products to the factory for refurbishment or reuse.
- Collective schemes for the recycling of products are attractive to manufacturers due to the shared risk and cost.
- Some manufacturers keep track of product installation by requesting the return of product commissioning documentation in return for a warranty.
- Provide disassembly instructions for refurbishers and recyclers.
- Reuse parts recovered from servicing where appropriate.
- Many products could be redesigned to be more easily dismantled for reuse and recycling.
- Incorporate recycled materials in products.

Promoting the reuse of building services

If a design team wants to specify reused or recycled building-services equipment, it will be vital to educate the client as to the capital costs and environmental benefits. It will also be vital to demonstrate that there will be no loss of building performance, that safety will not be compromised, and that the reused elements will not create a maintenance burden through poor reliability.

The arguments will need to be made very early during client briefing. Some clients and architects will be more receptive than others, particularly those familiar with refurbishment projects, and where the aim is to preserve much of the original building, for example a property deemed to part of a nation's heritage. In these instances, there is more likely to be a ready acceptance and enthusiasm for reuse that the building-services design team can build upon.

For new build, where the latest architectural features and fashions and engineering technologies tend to be uppermost in the client's mind (and therefore big selling points for those bidding to be on the design team), the option to reuse existing components and materials will probably be counter-intuitive to the client, if not also to the designers. In these situations the proposal to reuse components and materials will need to be presented differently. For example, the environmental arguments should be stressed to the client, particularly if they want to advertise their environmental credentials to their customers. On specific items, such as diesel generators, it will be relatively easy to show capital cost savings that can be used to discount the cost of items that the client would like to have in the building but is struggling to justify. In this respect it should be easier to enlist the support of the architect and interior designer.

For smaller, less costly products, such as motors, cable trays and pipes, the designer will need to demonstrate compound savings. This probably needs to be assembled prior to any project bid, and probably with the help of a reputable cost-consultant. The costs will probably be challenged, so they need to be detailed, accurate and accompanied by explanatory narrative expressed in clear, non-technical language.

Potential technical hurdles may need to be overcome – particularly concerns about efficiency, performance, maintenance and reliability. People with responsibility for these issues are likely to be nervous of reused equipment or recycled-content products. They will see themselves as inheriting the drawbacks and not necessarily enjoying the benefits. For these people the specifying process needs to contain accurate data on cost, maintenance requirements, equipment reliability and ease of obtaining replacement parts.

The interface issues between new and reused equipment will also need to be addressed carefully. There may be technical requirements to be satisfied, such as special connecting spigots or distance pieces, and warranty clauses that need to be investigated. Engineers need to discover whether the warranty of a new item of equipment is likely to be infringed or declared invalid by dint of its connection to a reused component.

Specifying reclaimed services is cost-effective when existing equipment is reused during a refurbishment, since the costs are solely for decommissioning and recommissioning the equipment. Some equipment, particularly electrical services, will require testing to ensure the systems meet prevailing safety standards. If a design team intends to specify reused equipment or recycled products from an outside source, the specifiers will need to ensure that the equipment will be of the required quality, which may require some form of formal quality control. While the capital costs of reused components may be lower compared to new components, the acts of finding, reconditioning and checking equipment will take longer and hence may cost more.

Every cost argument covering procurement, maintenance and relationship to the client's business objectives will need to be presented in detail. For projects where tenants' needs are not known, cost arguments will need to be presented for a suitable number of likely tenant scenarios.

The proposal need not be long and certainly not arcane. But it will need to be thorough, financially sound and easy to read. Unless the proposal is able to counter the views of those further down an organization who are risk averse – by instinct or through job function – it will be all too easy for the recycling argument to be derailed, no matter how supportive the client or architect may have been initially.

Even when the argument to reuse building services is won, the environmental benefits and cost savings will only accrue by ensuring the services contractors are onside. On many projects, waste management can be a major component of the contract. The same approach needs to be used for the recycling and reuse of building services. When the arguments are won at the client briefing stage, the services designer needs to identify the scope of reusing building services, and identify those on the project's critical path.

If procurement time is likely to be a factor, the design team and client should opt for early appointment of the relevant supplier or contractor. If the time issues of using recycled or reused services are not identified very early in the project, the design team may eventually be forced to buy new equipment – despite all the previous effort – just to keep the project on schedule.

Key design and specification issues for building services

In addition to the general matters affecting reuse and recycling discussed in Chapters 1 and 3, building-services plant and equipment are influenced by a number of other issues.

Design parameters

Building-services equipment is designed very precisely to suit the needs of a particular building. This will affect the capacities of all the goods installed, whether the working machinery or the means of distributing various fluids and electricity into, within and out of the building. Similarly, the internal climate of a building – the temperature, humidity and air quality that the building services help to create and maintain – depends on both the function of the building and spaces within it, and their geometry. Finally, the performance that the building services must achieve depend on the location of the building – both geographically, in terms of temperature, sun, rainfall and prevailing wind conditions, and in terms of the local infrastructure supplying water, gas and electricity and carrying away waste fluids for treatment.

In addition to the loads that the various building services equipment must meet, most of the equipment has a finite in-service life. While it is often claimed that a steel-frame structure is designed for a 30- or 60-year life, no one expects that it will suddenly cease to function at the end of these periods. This is not the case for a boiler or lift motor – such goods do, literally, wear out and need to be reconditioned or replaced. Different pieces of working equipment have different design lives.

In consequence, when proposing to reuse building-services equipment, it will always be essential to establish the length of the design life that remains before refurbishment or replacement is required.

For all types of services in a building acquired by a new owner, it will first be necessary to establish exactly what equipment is installed and its capacity – both overcapacity and undercapacity – is likely to be an issue. It will also be necessary to assess its condition, state of repair and performance. Such work will need to be undertaken by a specialist contractor with knowledge of the system in general, and with the type (manufacturer) of equipment itself. It will then be necessary to compare the capacity and performance with the operational performance needed for the proposed building use.

The components of building services

The vast majority of building-services goods are supplied by manufacturers in standard sizes or capacities. The task of the building-services engineer is to devise a system of parts that will meet the performance demands of a building. This has two important consequences:

1. It is extremely unlikely that an entire building-services installation (e.g. a heating system) would be removed from one building and reused in another.
2. It is highly likely that most of the components of a building-services installation could be reused to create a new installation in another building.

Some building-services items, such as air-handling units, are often specified as 'specials', usually meaning an otherwise standard component has been heavily modified to overcome a site-specific problem, such as constraints on space or requirement for higher output. A single air-handling unit, for example, may differ from its neighbours by having a larger fan to cater for a high local cooling load. Such units may perform at less than their peak efficiency when reused in a system with a more modest cooling load.

Full audits of the capacities and characteristics of components to be reused must be carried out to ascertain their suitability for use in new contexts.

Differences between imperial and metric dimensions will need to be identified, as will differences in the supply of incoming services (electrical voltage and frequency, gas and water pressure).

Legislation and regulations

Building services are subject to rather more regulation and legislation than many other parts of buildings, partly because of the dangers inherent in using electricity and gas, and partly because of the potential dangers to health if the air and water used by humans become carriers of disease.

Recently a third set of regulations has begun to emerge that is instrumental in reducing our use of Earth's non-renewable or scarce resources, in particular, energy and water. Such regulations and legislation vary from country to country and are being constantly updated, which has important consequences for reuse:

- It is unlikely that components from one country or regional regulatory authority (such as the European Union) can easily be reused in another.
- It is likely that some types of component removed from a building will no longer meet prevailing legislation or regulations such as energy efficiency, carbon emissions and health and safety issues.

A full list of prevailing regulations and EC directives is given in BSRIA *Directory Building Services Legislation: A directory of UK and EC regulations* (BSRIA Directory D10/2004 6th edition).

Changes to Part L of the UK building regulations in January 2006 will force the use of higher efficiency boilers (effectively condensing boilers) in most situations. In Europe, condensing technology has been the norm for many years.

Operation and maintenance manuals

Operation and maintenance (O&M) manuals are usually required for new equipment and newly installed systems. O&M manuals for reused equipment should be issued as for new equipment. It should be considered on a case-by-case basis whether the reused equipment will require more maintenance than new equipment, as this could adversely affect life-cycle costing and conflict with a client's requirement for reliability (or expectation of reliability). Where possible, full risk assessments of all building-services plant (including any reused items) should be measured against the client's business requirements.

O&M manuals may also be required to include a method statement and risk assessments for installing and maintaining equipment, provided by a building contractor. These would usually need to be developed in conjunction with the local heath and safety officer. In the UK this is required by 'Managing Construction for Health and Safety' – part of the construction, design and management (CDM) regulations.

Warranties and liabilities

For refurbished and reconditioned products, the engineering refurbishment company will be liable and should be asked to provide a warranty. For refurbishment projects (i.e. entire building-services systems), the mechanical and electrical contractor will be responsible for testing equipment to ensure compliance with health and safety legislation. There will be a liability on the design engineer to ensure that compliance is completed satisfactorily. For recycled-content products, the manufacturer will usually provide a guarantee.

Individual components in building-services systems have a specific design life. When reused, the life remaining will be less than for new equipment, and almost certainly different to adjacent equipment to which it is fitted to form a system. It should therefore be considered whether the remaining life is acceptable from both a elemental basis and a system basis. For example, expensive access may be required just for the one reused component in a system that might otherwise require very little or no maintenance over a very long period. Lifetime can be prolonged through refurbishment.

A relatively new solution to the issue of liability, which particularly suits the use of reconditioned products, is the provision of whole-project insurance under a single policy. This avoids the problems of faults or underperformance falling between the liabilities of the various different parties involved.

Reuse, reclamation and recycling options

Reuse in situ

When acquiring an existing building a new owner will need to establish carefully the condition of the building services. At a basic level this may be the work of a building surveyor, but it would also be wise to have an assessment of the various services by contractors qualified in undertaking technical performance assessments of services, and also able to undertake testing, cleaning, repairs and recommissioning of all components and the entire building-services system.

In the event that such an assessment highlights deficiencies in the system, there will be the opportunity to reuse as much of the installed equipment as possible. It will also generally be possible to repair, recondition, renew and perhaps upgrade any item found to be faulty or underperforming.

In undertaking such work it will be important to establish clear divisions of responsibility between several types of firm involved in the different activities for:

- assessment of the condition and performance of each entire building system;
- assessment of the condition and performance of individual items of plant and equipment, the remaining life of the item, and the opportunities for reconditioning the items;
- assessment of the demands on the building-services systems of the new occupier of the building or new use for the building, and design of new systems to meet these demands, using components of the installed system wherever possible;
- installation of reconditioned or new items of equipment and testing and recommissioning of the entire building system.

Reuse of salvaged or reconditioned products and reclaimed materials

This heading covers all working plant and machinery that is repaired and maintained as a matter of course. This includes items bought from suppliers of second-hand, refurbished or reconditioned machinery. The degree of attention that such items receive varies a great deal from none (electrical transformers do not wear out) to being returned to a factory, maybe the original manufacturer of the goods, stripped down, worn parts replaced, rebuilt and sold as a reconditioned item.

Quality expectations

The level of acceptable quality of the reused component will depend on the intended use. A client (or qualified representative) should be asked whether a lower quality than for new equipment is acceptable. A judgement should then be made on whether the quality will be adequate for the intended purpose and the duration of use.

Performance and safety tests to be undertaken

Before being reused, building-services equipment should be tested by the authorized supplier and contractor to ensure adequate operation and compliance with the regulations and standards that apply to the component or material in question, including safety requirements. The nature and number of tests will vary widely from material to material and component to component. The manufacturer's product information should be used as an initial guide.

Re-certification and product liability

Reused equipment should be tested by the contractor for contamination and safety. Responsibility usually lies with the contractor, unless the product or component has been refurbished in a factory, in which case responsibility will usually lie with the refurbishing contractor. Products with recycled content will normally be re-certified by the manufacturer.

Checks should be made to ensure that all parts of a reused or recycled item conform to the latest regulations and standards.

Recycled-content building products

RCBPs embrace only those products that are manufactured with some recycled material (such as plastic pipes) and exclude working plant and equipment that has been repaired or reconditioned and fitted with replacement parts. Plastic pipes and other products for public health services, water pipes and guttering are widely available with recycled content. Water tanks made partially or wholly from recycled materials are available. Refurbished boilers are available for reuse, but quantities are limited.

At present, however, plastics-based systems do not generally use recycled content, as it is widely felt that any impurity in the recycled plastics could impair their quality. Although this is often not the case, pipework manufacturers are sometimes unwilling to guarantee such products.

Mechanical heating/cooling/refrigeration systems

Plant used for heating, ventilation and air-conditioning (HVAC) has the greatest potential for reuse. Even in the absence of recycling legislation, HVAC equipment is probably reused more than any other type of building service. Much of this reuse is a consequence of building operation and maintenance strategies, where motors for air-supply fans are rewound as a matter of course, and high capital cost items like fan-coil units repositioned to suit a redesigned office layout rather than thrown away. When it comes to new build, however, maintenance cannot be a driver for reuse, which means that new building services products are almost always specified.

Gaining a greater understanding of the reasons why new HVAC products are almost always specified for new build is the key to successful uptake of reused equipment. In a word, specifiers buy new products for reasons of certainty:

- in availability and delivery;
- in warranties;
- in compliance with prevailing standards and regulations;
- in performance and compatibility;
- that no products in a system have used up any of their service life.

Even if the first four items in this list can be satisfied, specifiers will still be wary that the reused product may fail or cause problems much sooner than will a new product – and, most worryingly, within the building's defects liability period. The true causes of faults with new HVAC plant are already difficult to pin down, without adding the variable of a reused product. For services engineers, it is the equivalent of buying a new car and being confident that it has zero mileage on the clock.

The means of ensuring product certainty for reused goods will need to involve two forms of protection for specifiers: third-party certification, and explicit statements accompanying the product that list the items that have been replaced or refurbished.

Heating systems

Although reducing the level of carbon emissions from heating systems and products is increasing in importance, building clients, designers and contractors have to give greater priority to the certainties listed in the previous section. Similarly, improving the energy efficiency of HVAC equipment makes a far greater impact on carbon emissions over the whole life of a building than reducing the energy consumed during product manufacture – the embodied energy. Nevertheless reconditioning and reuse are viable options for many products.

Reuse in situ

All heating systems are suitable for reuse in situ given the appropriate risk assessment and changes to suit a new load profile. Boilers are designed and warrantied for no more than 20 years and so, when a boiler becomes available for reuse, a potential specifier would be able to decide quickly whether the boiler's remaining life makes its reuse economic.

While designers tend to specify excess heat-raising capacity (in the form of modular boiler units), even units with low run-hours are rarely reused. Reuse is also hampered by legislation that periodically raises the bar on energy efficiency. That said, boilers are made up of many parts, including burners, valves, controls, insulation materials, tubes and fans, all of which can be replaced at the end of their life. Larger boilers offer greater opportunities for in situ refurbishment to extend life because of the greater proportion of the materials and cost in the large mass items. Combined heat and power units can also be reused in situ and their life prolonged by on site maintenance, repair and recommissioning.

Underfloor heating and cooling is achieved using either electric- and water-based systems. Water-based systems generally use plastic tubing (polyethylene, polypropylene or polybutylene) or copper pipe laid as a serpentine coil on (or within) a floor construction. Floor heating pipes are often quoted as having a design life exceeding 50 years, guaranteed by the manufacturer – a life that is comparable with that of the building itself.

Reuse of salvaged or reconditioned products and reclaimed materials

As the wearing parts of boilers can be replaced, and other parts can be cleaned, there is already a healthy market for reconditioned boilers. There are also many firms offering boilers for hire, since the owners can be confident of being able to recondition the products from time to time to raise their value again.

Combined heat and power (CHP) units are designed as building-specific items to match the precise demand loads. Reusing a CHP unit in another location would require electrical and heat loads similar to that of the original installation and, therefore, is unlikely. It could nevertheless be possible to find individual components from the reconditioned market.

The ease of reuse of underfloor heating and cooling systems will depend on the floor construction. The extraction of pipe embedded within a concrete screed will be difficult without some damage being inflicted on the pipe. The pipe would also need to be cleaned, pressure tested and rewound before it could be reused. Note that plastic underfloor heating pipe work may become hardened and embrittled after prolonged use, and therefore very difficult to extract, rewind and relay without breaking.

Figure 8.1 Cast-iron radiators stored in an architectural salvage yard

Source: Buro Happold

Systems where the pipe or electric heating elements are laid directly under floorboards, preferably clipped to preformed insulation panels, will be easier to demount and reuse. Here is a good example where designing for ease of recycling or reuse could be a strong candidate for sustainable design.

Recycled-content building products

The most likely opportunities for using RCBPs in heating systems are various plastic products, for example plastic pipes for water-based underfloor heating and cooling systems.

Refrigeration equipment

Reuse in situ

All types of refrigeration system are suitable for reuse in situ given the appropriate risk assessment and changes made to suit the new load profile.

It should be established whether a refrigerant in a product being assessed for reuse is on the EC's banned list (European Directive 2037/2000) and, if so, whether it will need to be replaced with a more environmentally friendly refrigerant within the design life of the (reused) air-conditioning device.

Reuse of salvaged or reconditioned products and reclaimed materials

There is a well-established market for reconditioned refrigeration plant of all sizes. As with reuse in situ, it is likely that old plant will contain a refrigerant that is now banned or, at least, discouraged because of its capacity for causing the depletion of ozone. Firms reconditioning refrigeration equipment will usually replace such refrigerants, if not as a matter of course, then on demand.

Reconditioned components such as valves, pumps or filters are also available, though the number of sources is limited.

Recycled-content building products

The most likely opportunities for using RCBPs in heating systems are some plastic products.

Ventilation and air-conditioning systems

Ventilation and air-conditioning equipment is probably reused more than any other type of building service. Much of this reuse is a consequence of building operation and maintenance strategies, where motors for air-supply fans are rewound as a matter of course, and high capital cost items like fan-coil units repositioned to suit a redesigned office layout rather than thrown away.

Reuse in situ

During building refurbishment, storage and distribution equipment can be reused subject to its condition and load requirements, and its ability to interface with new components (such as reused ductwork with new volume-control dampers and fire dampers, and vice versa).

Ventilation systems

A complete ventilation system would only be reused in situ. The condition and duty should be adequate for the intended purpose and load requirements, including meeting recommended minimum ventilation rates, volume flow and velocity, noise requirements and specific fan power. Depending on the design life or remaining life of the separate components, items such as fan and pump motors can be refurbished and rewound. The criteria listed in the introduction to this section would also need to be satisfied. The act of refurbishing these items may force the need for other items to be replaced or repaired, such as filters and casings.

Ductwork systems

Distribution systems comprising pipework and ductwork are generally building-specific. If, during a refurbishment, the systems in the building are to remain essentially the same, with no significant changes to load requirements or air- and water-flow characteristics, then the distribution system can be reused. However, if there is a partial or total redesign of the building's services, reuse of the existing distribution system may be more problematic. The existing system would need to be audited, broken down to its constituent parts, and a view taken on which elements are suitable (and economic) for reuse. These would normally be long runs of pipework or high capital-cost assemblies of valves and strainers.

Ductwork is difficult to demount and reuse as it can be easily damaged. It may also be difficult to obtain integrity to the requirements of the building regulations (as specified in BS 5422) once the ductwork joints have been broken. Again, long runs of primary ductwork are likely to prove more economic to reuse than secondary ductwork, and particularly small and more fragile flexible ducts. As ductwork is mostly made of metal, recycling is probably the most sustainable option. Designers should always enquire whether a ductwork contractor is prepared to use recycled materials, and specify accordingly.

The casing for air-handling units can be reused in situ, but many of the components, such as fans, coils and humidifiers will usually be replaced with new equipment. It is possible to refurbish these units, but the cost of disassembly, transportation back to the factory, refurbishment, testing and storage will cost more than a new unit.

Control devices

Motor control centres and small local controls can be reused in situ, but will not generally be appropriate for reuse on a new site. Obsolescence is a major issue here. Controls are one of those building services subject to constant improvement in performance and functionality. Older units are more likely to be dependent on bespoke software that may not be able to communicate with a new building-management system. Even if the controls are not deemed obsolete, the original manufacturer(s) should be contacted to establish the longevity and upward compatibility of the product.

Reuse of salvaged or reconditioned products and reclaimed materials

Direct-expansion (DX) *air-conditioning units*

As they are effectively shrink-wrapped, modular, factory-made items, direct-expansion (DX) air-conditioning units are eminently suitable for reuse. There are various issues that would need to be addressed, such as recommissioning and recharging with a refrigerant with a lower ozone-depletion potential that may be chemically different to that originally used. The insulation on the pipework of air-conditioning plant may also contain CFCs or HCFCs. These are banned substances that may preclude reuse. There is also no industry dealing in reusable air-conditioning units, so availability is limited.

Pumps and fans

The duty, condition and design life of a pump can usually be easily established. The nature of the application will also determine whether a pump can be reused. For example, pumps should only be reused with the same fluid operating under the same pressure and temperature conditions, unless the refurbishing manufacturer or contractor is prepared to warrant the performance of the pump under the different conditions. For potable water applications the designer will need to check whether the water undertaker needs to be notified of any pump and booster installation, as specified in water regulations.

Fans should be of adequate condition for the intended purpose, and it should be established whether there is any damage to blade and foils. The duty should be adequate for the new purpose and load requirements, and the size should match the ductwork.

Terminal devices

Fan-coil units are reusable within their normal design life and are relatively easy to move during full or partial refurbishment, particularly when there is a change from cellular spaces to open plan and vice versa. Minor cosmetic damage would generally be acceptable as ceiling-mounted components are not usually visible. Perimeter fan coils are usually housed in an architectural casing, which will probably be renewed for aesthetic reasons.

Care should be taken during removal and reinstatement, particularly with such items as heat exchangers, the cooling fins of which are easily damaged to the detriment of subsequent performance. The design capacity should be checked to ensure the heat exchanger is adequate for its intended reuse.

Testing

In the UK all reused HVAC plant should be tested by the contractor to ensure adequate operation and compliance with relevant Approved Documents in the UK building regulations. Heating systems will need to comply with Approved Document L2, and adequate ventilation for combustion appliances to Approved Document J. The latter document also addresses the issue of reusing flues. Reused ventilation devices will need to conform to Approved Document F. Recommended design criteria for HVAC installations are published in the UK (CIBSE, 1999) and guidance for testing ductwork leakage is also available (HVCA, 2000). Each country has its own set of similar regulations and design guidance.

Recycled-content building products

The most likely opportunities for using RCBPs in heating systems are various plastic products.

Piped supply and disposal systems

General considerations

The supply, storage and distribution of liquids and gases include the following:

- water supplies;
- steam systems;
- gas systems;
- liquid fuel systems;
- special liquids systems (a specialist area not covered);
- fixed fire-suppression equipment.

Various materials are used for pipework serving the uses listed above, such as steel, plastic and copper. The pipework sector benefits from a relatively mature recycling industry. Not only can it collect and process waste piping materials, but it can also provide reconstituted products for both new build and refurbishment projects. There is also a growing industry in recycling plastics, particularly for use in water and public health installations. Some products are available that are tested, third-party certified and guaranteed. A growing number of products also have the backing of insurance companies.

General design considerations

The main issues of reusing or recycling gas and liquid supply, storage and distribution systems are:

- design loading and capacity;
- design and remaining life;
- condition including corrosion and contamination;
- maintainability;
- legislative and health and safety requirements.

Reuse in situ

During refurbishment, storage and distribution equipment can be reused dependent on the condition of the equipment and the loads to which it will be subjected.

During refurbishment of a building's services, the distribution systems can be reused depending on the new design for the building and services with regard to duty requirements, the remaining life of the services, corrosion and contamination. Cleaning should be completed and any physically damaged sections should be replaced to maintain integrity.

There is some reuse of water storage tanks that can be refurbished and reused in situ. The practicality of reuse will depend on the condition and degree of contamination discovered.

Testing

Equipment refurbished and reused in situ should be tested by the contractor to ensure adequate operation and compliance with prevailing legislation. Biological tests should be completed on water supply systems, and safety tests should be undertaken on the steam, gas and liquid fuel systems. Responsibility for refurbished equipment will lie with the refurbishing contractor.

Reuse of salvaged or reconditioned products and reclaimed materials

Pipes and fittings generally come in standard sizes that increase the opportunities for their reuse. However, pipes tend to be cut and bent to local requirements, and this might limit reuse. Long straight runs are likely to prove more economical to reuse than short, complicated sections.

Storage equipment can be obtained for reuse, but limited sources and a lack of advertising make this option cost- and labour-intensive.

Certain manufacturers are able to refurbish old tanks or recycle them by using sections to make up new tanks. This service is generally not advertised and tends to be used by building owners as part of their operation and maintenance activities. However, the service is equally valid for new build and refurbishment projects.

Minor cosmetic damage would generally be acceptable as distribution elements are not highly visible. Again corrosion and contamination are issues that need to be resolved before reuse.

Testing

Reconditioned equipment will need to be tested by the re-manufacturer to ensure adequate operation and compliance with prevailing legislation. Biological tests should be completed on water supply systems, and safety tests should be undertaken on the steam, gas and liquid fuel systems. A system making use of reconditioned equipment should be tested by the contractor for contamination and safety; responsibility will therefore lie with the contractor.

Recycled-content building products

A wide variety of plastic pipes made from recycled-content materials is available and they can be used in both new build and refurbishment projects. These can be found using one of the growing number of websites providing databases of sustainable building products, including RCBPs (see pp205 and 206).

Water supplies

Distribution systems

Distribution systems are generally designed to meet the specific needs of a building. If, during refurbishment, the systems in a building are to remain essentially the same, then the distribution system could be reused. This is providing there are no significant changes to load, flow or capacity requirements, that the same size of piping and storage is required, and that the condition of the systems is acceptable.

If there is any significant difference between an installed system and new requirements to be placed upon it, then the existing distribution system may need to be partially or completely replaced. The extent of the distribution system will also be a factor. Some parts of it may need to be removed, and new sections added.

It is possible that materials previously used for handling corrosive or harmful waste (such as swimming-pool water or services used in kitchens or toilets), or waste operating at extremes of temperature (such as hot and cold water services), may be suitable for reuse in a less onerous environment. However, a full risk assessment should be carried out to determine whether the refurbished material or product presents a health or safety hazard in the context of its reuse. A risk assessment should also prove that the material has not been weakened to the extent that its remaining life is too short for its use to be economical.

Corrosion of any kind will limit reuse, but evidence of bacterial growth or deposits would also be a serious consideration, particularly for items such as valves and similar fittings. Adequate cleaning and disinfection should be carried out and biological tests completed before any part of a fluid distribution system can be considered for reuse.

Hot and cold water storage

The condition of hot water storage items needs to be good. Products should be checked for corrosion and contamination, and insulation checked for integrity and replaced if necessary. In the UK, the performance of the reused water storage would need to meet the current standards laid down in Part L of the building regulations. In other countries EU directives and national codes, standards and legislation will apply.

Pumping systems

Pumps are highly suitable for reuse. It will be necessary to check and test that the pumps will match the load profile of the new installation. The condition and design life of the reused pumps should also be ascertained through a risk assessment. Reconditioned pumps are available, and engineering companies specializing in pump refurbishment will normally provide a warranty.

Water supply pipework

If pipework is to be reused in situ, then it should be checked to ensure that it will meet the capacity and required flow rates. It should also be checked for corrosion and contamination. Pipe sizes and runs should be checked to ensure that they meet prevailing legislation, standards and codes of practice. Recycled pipework is readily available and the supplier will normally provide a warranty.

Valves for water supply

As with pipework, reused valves should be checked for operation and corrosion. They should also be checked to ensure they are the right fittings for any new pipework in the system. Refurbished valves are available, although they are not advertised.

Steam supply, storage and distribution

Steam systems for heating and cleaning purposes are common in many types of older building. In newer buildings, steam systems tend to be used only in industrial premises, hospitals and laboratories of various types. In most cases where such buildings have reached the end of their lives, it is reasonable to assume that the steam-based system will not be far behind. High temperatures and pressures, along with some inevitable corrosion, will probably result in components that are only fit for recycling rather than reuse.

However, the designer should check whether any of the components in the system have been renewed or refurbished to a level and within a timescale that would make reuse an acceptable option. It will be useful to study the system's maintenance history for any evidence of renewal.

Complete steam systems are unlikely to be suitable for reuse in a different building, as generally no two buildings are the same. Steam systems are also likely to be damaged during decommissioning and recommissioning. Some recycled-content products are available that will normally be guaranteed and warranted.

Gas supply, storage and distribution systems

Gas supply and storage systems will not suffer internal degradation, unless the gases are corrosive. It will be necessary to understand the relationship between the gases and the material used to transport them to understand whether any components are likely to be suitable for reuse. This will also apply to liquid fuel storage and tanks storing gases at atmospheric pressure. Pressurized storage vessels for gases such as fuel gas and liquid petroleum gas will need to be risk-assessed prior to reuse, whether in situ or at a new location.

As with other piped services, complete gas supply systems are unlikely to be suitable for reuse in a different building, as generally no two buildings are the same. Gas supply systems are also likely to be damaged during decommissioning and recommissioning. Recycled content components are available and the component parts will normally be quality guaranteed and warranted.

As with other kinds of pumps, gas supply pumps can be reused after a risk assessment has been carried out, their condition determined, and appropriate refurbishing has been carried out and warranted. Gas supply pumps are also suitable for recycling.

Fixed fire-suppression systems

Fixed fire-suppression systems in all countries have to conform to a variety of standards and regulations concerning both their operation and their safety when fluids operate at high pressures (above about 0.5 bar). There is usually nothing in these standards that would prevent reuse of fire-suppression components, but it will probably prove onerous to meet the legislative requirements using reused components, and may therefore not be cost-effective.

For in situ refurbishment projects, a fire-suppression system such as a sprinkler system can often be reused. This would not normally include the pipework as it will probably be corroded, but the dial pressure gauges and the sprinkler heads could be reused after suitable refurbishment.

When considering reuse of sprinklers, the specifier should establish that the type of sprinkler to be reused is appropriate for its new role, as different types of sprinkler are used for different levels of fire hazard.

Few buildings share identical dimensions, so complete systems are unlikely to be suitable for reuse. Systems are also likely to get damaged during disassembly and refitting. In this instance, it will probably prove uneconomical to try and reuse odd lengths of pipework, although certain components can be reused, such as pumps, valve sets and tanks, depending on their condition and the outcome of a risk assessment.

For water mist and fine water-based spray systems, the condition of the equipment to be reused will need to be good. The pump (if fitted) will need to maintain discharge pressure for the duration specified in fire protection regulations.

Recycled-content components are available that will normally be quality guaranteed and warrantied.

Foam systems are similar to fixed fire-suppression systems, except that foam is corrosive. On this basis there is little opportunity for reusing the pipework. However, a tank may be suitable for reuse if it has an inner skin. A tank supplier should be asked to assess the cost-benefit of having the tank refurbished.

In the UK, halon fire-suppression systems should all have been decommissioned by 31 December 2003, so it should be established that the system is not halon-based.

Disposal systems

Drainage, sewerage and refuse disposal incorporate a number of building services disposal systems including various forms of waste-handling equipment and associated pipework, such as:

- wet-waste handling products;
- solid-waste handling products;
- gaseous-waste handling products.

In all cases the materials and equipment used for handling these wastes need to conform to the relevant current standards and regulations associated with the disposal of waste. It is extremely unlikely that complete systems will be available for reuse, and individual items of plant and equipment should be considered separately.

It is possible that materials and equipment previously used for handling corrosive or harmful waste, such as swimming-pool equipment or services used in kitchens or toilets, or waste systems operating at extremes of temperature, such as hot and cold water services, may be suitable for reuse in a less onerous environment. However, a full risk assessment should be carried out to determine that the refurbished material or product will not present a health or safety hazard in the context of its reuse. A risk assessment should also prove that the material not been weakened to the extent that its remaining life is too short for its use to be economical.

Conversely, products or materials not designed to operate in an onerous environment should not be reused in that context, unless the responsible supplier or contractor can prove that the conditions to which the materials will be exposed will meet all the requirements.

A number of recycled-content products are available for use in drainage and other disposal systems. For example plastic pipework may be made from recycled plastic. Also disposal systems often require the use of in situ concrete that can be made using recycled aggregates. The formwork for such concrete can be made from 100 per cent recycled PVC. The formwork used to make cylindrical concrete drainage pipes from 100–1500mm in diameter can be made from cardboard tubes made from post-consumer paper fibre.

Electrical supply, power and lighting systems

Electric power and lighting services products include:

- power-storage devices;
- transformation devices;
- protection devices;
- treatment devices;
- measuring and recording devices (a specialist area not covered);
- distribution devices;
- terminal devices.

Reuse, reconditioning and recycling in electrical services

Reuse in situ

Electrical systems generally have a design life of between 20–30 years. Refurbishment within this period should allow an existing electrical system to be retained in whole or in part, as long as the condition is adequate and the new design load can be met by the existing system or by upgrading the existing system. In all cases the system must be safe and tested to the appropriate standard. The load centre should also be in a similar position in the new design to facilitate the reuse of existing equipment. For any new build, proximity to the load centre will largely determine whether the existing high-voltage (HV) equipment can continue to be used.

Reuse of salvaged or reconditioned products and reclaimed materials

Reused electrical equipment should be tested by the contractor for legislative compliance and for adequate operation, capacity and legislative compliance to the appropriate national standard (BS 7671:2001 in UK). For this reason responsibility shall lie with the contractor. Even where equipment has been refurbished, the contractor will still be responsible for ensuring that the electrical system, including the refurbished items, complies with legislative and health and safety requirements. Products with recycled content will normally be certified by the manufacturer.

Recycled-content building products

The most likely opportunities for using RCBPs is where recycled plastics can be used. This will generally be possible only for soft thermoplastics, such as cable insulation, that can be recycled. The hard plastics so widely used for their insulating properties in electrical equipment are usually thermosetting plastics that cannot be recycled.

Power generation, supply and distribution

Standby generators

Standby generators are a prime candidate for reuse because their life expectancy is long as a consequence of infrequent use. Such equipment is regularly refurbished with replacement batteries and controls. This not only prolongs the life of a generator in situ, but is a factor if reuse is being considered for a different site. Age will ultimately determine whether the generator can be reused. Generally speaking, the newer the generator, the more compact the size and the more efficient it will be at generating electricity. Both these factors will need to be traded against the cost of providing space – both for the generator and for its fuel-storage requirements.

Distribution and switchgear

Most of the components of distribution and switchgear are provided in standard sizes and this increases their potential for reuse. This applies to protection devices, relays, circuit-breakers and switchgear, as well as distribution boards, cabling, connectors, junction and terminal blocks, cable ducts and trunking.

High-voltage switchgear, circuit-breakers and transformers use sulphur hexafluoride for insulation and to inhibit erosion of contact breakers. This gas can be collected and recycled in reconditioned or new equipment. However, other substances and oils are now available that are considered more environmentally friendly than sulphur hexafluoride and this may persuade a client against its reuse.

Power storage devices

The operation and condition of rectifier/charger, inverter and batteries in uninterruptible power supplies (UPS) all need to be verified prior to reuse. So too the other electrical components including capacitors, relays, circuit-breakers and contactors. The controls will need to be compatible with the other elements of the installation.

Large batteries used by standby generators, for emergency lighting and in uninterruptible power supplies, can all be bought as reconditioned items.

Transformation devices

Transformers deteriorate less than most equipment installed in buildings and for this reason there is a mature market for second-hand/salvage goods. Prior to their reuse it is likely that old oil-cooled transformers will need to have the coolant replaced by a material that is more environmentally benign than some materials formerly used.

For transformation devices it is important to establish the loading, as dry transformers are less efficient at low loading. Location is also important as oil-filled transformers pose a fire hazard and may need automatic fire extinguishers if located internally.

Terminal devices

Motors

Motors are highly suitable for reuse and recycling. Large- and some medium-sized electric motors are regularly rewound and refurbished as part of a building's operation and maintenance routines. Conversely, rewound motors are difficult to source for new build. The motor duty and size needs to be matched to the equipment being driven and its application, such as a fan in an air-handling unit. This might be more difficult to achieve for in situ reuse than for new build, where the designer has more freedom to source components.

Second-hand and reconditioned motors are widely available, though at present they are not generally used for new buildings.

Luminaires

In many ways lighting and luminaires are one of the most important electrical terminal devices because of their embodied energy, their electrical performance, aesthetics, maintainability and electrical safety. They are also used in great numbers.

As with other critical services, lighting devices are regularly maintained, which means that – given a good audit trail – designers can identify fittings that have not yet reached the end of their service lives. It should also be relatively easy to assess which fittings conform to prevailing legislation and lighting guidance, thus enabling their reuse. That said, recycling the constituent materials of luminaires is a mature industry, so lamps are not normally reused. This market is being further stimulated in Europe following the adoption of the Waste Electrical and Electronic Equipment Regulations and Restriction of the Use of Certain Hazardous Substances Directive that cover much lighting equipment.

Reuse of luminaires is commonplace during general facilities management and during refurbishment. In principle, many fittings salvaged from a building would be suitable for reuse, subject to the need to test the products. This may be viable and cost-effective for high-value luminaires. For new construction the financial benefits of specifying new products would generally outweigh the environmental benefits of reusing salvaged luminaires.

If a designer does opt to reuse lamps, for example for an in situ refurbishment, then the fittings would need to be assessed for compatibility with prevailing recommendations. The colour rendering and efficacy will need to be adequate to the task and glare should be eliminated, as per current legislation. Compatibility with new luminaires should also be established.

Luminaires tend to be thrown away after their first life. They are often light and fragile items and easily damaged when being removed. Extreme care would need to be taken during de-mounting, transportation and their reassembly in a new location. This, more than anything, will probably determine the likelihood of their reuse, either in situ or in a new location.

Switches and sockets

Small items such as switches and wall sockets are not economical for anything other than in situ reuse, as they would require electrical testing after salvage before they could be reused. This would be very labour-intensive and uneconomical for such a low-cost item.

Information and communications products

Information and communications services products addressed in this section include building management systems (BMS), safety and security information systems and communications cables.

Control systems comprising BMS generally have a life expectancy of 10–15 years. However, control systems evolve rapidly, quickly leading to technological obsolescence within two to five years. There is also little likelihood for reusing control components due to issues of compatibility between products and sub-systems. In many cases reuse will be limited to in situ refurbishment projects, covering both software upgrades and additional hardware.

The recycling of electronic parts is governed by the Waste Electrical and Electronic Equipment Regulations and the Restriction of the Use of Certain Hazardous Substances Directive. Controls cabling is sometimes reused as part of the upgrade or refurbishment of the control system, and could be reused in a new build project once the technical requirements have been satisfied and the needs of the risk assessment have been met.

The degree to which communication cables can be reused will depend on the control strategy. Control systems and software have limited life due to rapid technical advancements. The client may also desire fast data-transfer rates, and older cabling may be unable to deliver the required speed. That said, interface products are available that can enable communication between otherwise incompatible systems. A system risk assessment should draw distinctions between critical and non-critical cabling requirements, and attempt to make the case for reusing cabling for the non-critical applications.

As with other electrical building services, all reused components and equipment will need to be tested by the contractor to ensure adequate operation, functionality and legislative compliance.

The work of identifying reuse opportunities, testing, installing and commissioning information and communication systems would be undertaken entirely by specialist subcontractors. While there are undoubtedly many opportunities for reusing both components and cabling for information systems, it would probably be difficult for the project team of a reuse/recycling building to persuade the subcontractor to reuse goods because there would be little or no benefit to them to do so. Indeed, they would be required to undertake more work and to take on responsibility for warrantying the reclaimed goods. When this additional effort is balanced against the small environmental benefits (given the low mass of materials involved) it is unlikely that reuse and recycling in this area would be viable, except, that is, in the case of reuse in situ when an existing system upgraded.

Safety and security information systems

Safety and security information systems tend to be stand-alone systems, installed by a specialist contractor. Many of the separate components should be reusable, although the usual provisos on hardware and obsolescence for items like cameras, swipe-card readers, video screens and computers will apply. Designers should be aware that otherwise innocuous software upgrades can have serious ramifications for the reuse of hardware that outwardly appears serviceable. For example, security communications software written for operating systems of the 1990s may not be available for more recent systems, forcing the continued use of the older operating system. Having decided to use an older operating system, the client may subsequently find the decision will limit the functionality of any new software applications, such as those based on internet protocols.

Conversely, a decision to run the latest operating system on reused front-end computers may force memory upgrades on the computers – and on other items in the system – that the designer hoped to reuse without any changes. This hardware may no longer be available, or very difficult to source. Furthermore, older security products may not have Windows plug-and-play devices, and software may need to be rewritten so that the new operating system recognizes them. This would cover controllers, modems and access devices. The continuing availability of cards or tags should also be established.

To make matters worse, compatibility (and reliability) is very difficult to ascertain until all systems are installed and tested. Ideally, a specialist contractor should be asked to carry out the risk assessment and be asked to identify hidden costs.

Note that the recycling industry for electrical systems is maturing, and therefore the recycling of obsolete computerized systems, as opposed to reuse, may be the best approach.

Access controls

As with security systems, access controls tend to be stand-alone systems. Online system components can be reused, though cabling would be replaced. Components, controllers, modems and access devices will need to be tested for compatibility. The ready availability of replacement cards or tags should also be established.

Presence detection

The level of detection required by the client should be determined before any decision is taken to reuse presence-detection equipment for new buildings or in situ refurbishment. Small items such as mechanical switches and magnetic contacts would be more economical to buy new. The diodes in photoelectric sensors have a limited life so would generally not be reused. Passive infrared sensors are affected by heat so their operation should be verified prior to reuse.

Fire- and smoke-detection systems

Components of systems, control panels, repeaters, sounders, detectors and call points can all be reused in a compatible system. For the reuse of cabling it would be necessary to bear in mind the points made in the section on electrical building services. However, recycled-content cable suitable for use in fire- and smoke-detection systems may not be available due to the fire-rating standards that apply to fire cables.

Communication cables and closed-circuit television systems

All cables destined for reuse should be undamaged and the reuse of cabling would require consideration of the points made in the section on electrical building services. If electromagnetic interference will be an issue in a new context, screened or shielded cable may be required.

If it is decided to reuse components of CCTV systems, the needs of the client will have to be matched to the specification of the equipment to be reused. This will include the required balance between resolution and colour, as monochrome systems have higher resolution, as well as the likely need for screened cable.

Opportunities to specify RCBPs in control and communications systems are likely to be limited to the plastic coverings of cables.

Lifts and escalators

Lift cars and the associated lifting machinery are generally reused during a major building refurbishment, depending on the age of components related to their design lives. Lifts also have many component parts that can be individually refurbished or replaced by reconditioned goods. For example, lift motors can be refurbished or replaced separately.

There is little evidence of lifts being removed and reused at new locations, although this does not preclude the possibility. Older lift installations tend to be tailored to the precise form, construction and dimensions of the shaft in which the lift cars are destined to run. The rails are not only fitted directly to the shaft, but the tricky process of aligning the rails and the lift cars to the floors is also very site-specific.

Lift design and installation tends to be a turnkey contract carried out either by a lift supplier or a nominated installation contractor. As a consequence there is little general knowledge among the building-services profession about the minutiae of lift systems. On the other hand, the lift companies will be a rich source of information and guidance.

In the late 1990s, lift design moved away from guide rails fitted directly into shafts to self-supporting aluminium guide rails, braced within the shaft. Many lift designs also became modular, enabling the equipment to be lifted section by section into the building core along with the factory-assembled lift car. Rather than a lift installation being built over a number of months, the whole modular assembly could be installed in just a few days.

The modular, bolt-together nature of these more recent lifts should mean they are more able to be de-mounted and used elsewhere. The newer systems also tend to come in a range of standard sizes, which would aid reuse. It is therefore possible that a new lift installation could be a mixture of new and reused elements.

As lift cars are a mixture of heavy engineering and lightweight architectural finishes, the services engineer and architect need to work together to maximize the use of reused and recycled materials and components. The visible finishes in the lift car will almost always be replaced, and the architect should be encouraged to obtain materials from sustainable sources.

The key issues that need to be addressed by the design team are:

- design loading and passenger capacity;
- design and remaining life;
- condition, particularly with regard to safety;
- maintainability;
- legislative and health and safety requirements;
- aesthetics.

There is no industry for reclamation, refurbishment and resale of transport services equipment, but it is readily refurbished in situ, including replacing car, panels, cables and motors, as required.

Escalators and conveyors are similar to lifts in that they would be refurbished in situ, with the replacement of some components, as necessary.

All reused equipment needs to be tested by the specialist contractor and therefore responsibility will lie with the contractor, unless being refurbished in situ, in which case responsibility shall lie with the refurbishing contractor.

APPENDICES

A The Practice and Management of Demolition Activities

Recent changes in the demolition industry

While this book is not intended as a guide to demolishing buildings, some understanding of the processes involved gives a valuable insight into how building components and materials are likely to find their way into the reuse and recycling market place. It will also highlight the need to influence normal demolition practices if equipment, products and materials are to be salvaged for use in constructing another building.

- Demolition activities and processes are undertaken in accordance with codes of practice that ensure they are undertaken in a responsible and safe manner (in the UK, BS 6187:2000, the Code of Practice for Demolition).
- Segregation of materials is essential to ensure the maximum benefits from the various materials streams produced by the demolition process.
- The type and method of demolition employed is dependent on:
  - the location of the building;
  - the type of structure, construction and materials involved;
  - the space available on the site for segregation and storage;
  - how materials are to be disposed of;
  - safety of operatives undertaking demolition work;
  - the time and money available to undertake demolition.
- Most valuable equipment and components with potential for reuse are removed during the 'soft-strip' process.
- The careful deconstruction of buildings usually takes longer than the demolition of a building.
- A cost–benefit analysis will help ascertain the true cost of different means of disposing of materials and the opportunities to maximize their potential for recycling.

Until the 1970s the demolition of buildings was still largely done by hand using lightweight equipment, apart from the final stages when ball and chain was used to demolish concrete and masonry construction. In consequence, many items were removed from buildings with some care and little damage. Since that time, the demolition industry has been subject to a number of major influences and changes:

- pressure to reduce the costs of demolition and materials disposal;
- pressure to complete demolition activities in ever shorter times;
- pressure to improve levels of health and safety;
- demolition is no longer viewed as just 'demolition' but as a material recovery and disposal activity;
- a greater concern to protect the environment;
- a move to embrace the 'whole life cycle' concept of materials;
- a developing global market place for the demolition industry.

Many of these changes have resulted from changing legislation. In the UK changes have occurred in:

- Health and Safety legislation;
- Construction (Design and Management) (CDM) Regulations;
- Management, Health, Safety and Welfare Regulations;
- Construction Health, Safety and Welfare Regulations;
- Control of Substances Hazardous to Health (COSHH) Regulations;
- Asbestos Regulations.

In response to these various influences, demolition practices have changed to embrace:

- a move away from the heavy use of manpower towards the increasing use of remote working with 'one man and a machine';
- the development and introduction of new plant and equipment incorporating, for example, 'super long reach' and 'robotic and remote-controlled' techniques;
- less deconstruction of buildings and more demolition, and hence fewer building components removed without damage.

Current demolition practices

It is always important to ensure that all materials streams resulting from demolition are carefully managed. In order to increase their potential for reuse or recycling, materials and components need to be segregated and stored, either on site or in a nearby temporary location, prior to subsequent handling. During this stage it is also important that contamination or other damage is prevented in order to preserve the potential for reuse or recycling and, hence, the value of the products. This is especially true for the higher-value products of the soft-strip stage.

The main drivers for selecting methods for soft strip and demolition of a building, as well as the disposal of the materials streams, are commercial. Commercially valuable components and materials with potential for reuse or recycling need to be identified in advance by means of a predemolition building audit. This will have a large influence on how the soft strip and demolition is conducted, for instance whether selectively and with great care, or with less concern for avoiding damage to components.

Current practices employed by the demolition industry are set down in codes of demolition practice, for example in the UK, BS 6187: 2000, The Code of Practice for Demolition. In addition there are guides to good practice such as the Institution of Civil Engineers' Demolition Protocol (ICE, 2004), which goes beyond a code of practice in recommending in great detail how demolition materials should be dealt with to minimize the quantities of waste that are sent to landfill sites.

Planning for demolition

The following initial activities need be undertaken by a contractor preparing to undertake demolition:

- desk study to record all available information;
- completion of and access to decommissioning reports;
- detailed on-site surveys, possibly including a predemolition audit;
- identification of all structural features and all possible hazards;
- consideration of pertinent and relevant legislative requirements;
- proposal of safe working arrangements;
- selection of suitable demolition principles and methodology.

The completion of the above tasks will provide the information required to establish the hazards, risks and opportunities related to health and safety and environmental issues, structural composition, material recovery opportunities, and disposal of hazardous or unusable waste material (e.g. asbestos) and thus ensure the safe, economic and efficient management of the demolition process.

The management and demolition techniques to be used can then be planned in relation to these hazards to minimize the risks. This includes the use and effective implementation of exclusion zones, environmental protection methods, material recovery options and mitigation of the identified potential risks during the work.

Predemolition audit

The value of a building due for demolition depends on what goods and materials are in the building. A range of organizations conduct audit services that establish just what goods and materials are in a building. Such an audit would establish the following:

- how the building is constructed, and hence how it can best be deconstructed and demolished;
- equipment and other valuable items that can

be removed in the soft strip;

- approximate quantities of other items and materials with salvage value;
- approximate quantities of items and materials with no salvage value.

The success of the audit process will be increased if reliable information is available – drawings, method statements, bill of quantities and inventory of equipment, specifications, and so on. With such information it will be possible to plan the deconstruction process more effectively thereby:

- ensuring suitable and safe access to building elements and fixings with minimal machinery requirements;
- ensuring toxic materials are dealt with appropriately;
- planning the sequence of dismantling operations to minimize deconstruction time;
- planning deconstruction and demolition to ensure the separation of materials/waste streams that is essential for economic recovery, handling and recycling.

Soft strip or stripping out

The soft strip of a building is a key stage in all forms of demolition, whether during refurbishment, partial demolition when part of a main structure or structures of a development site are retained, or total demolition. It is usually the first stage of any demolition project after the initial planning stages of the process have been completed and services disconnected. It may precede or immediately follow the removal of any asbestos. Soft strip may form part of the main demolition contract or be let as a separate package.

This stage of the demolition process is generally labour-intensive and provides important opportunities for recovering components and materials for reuse or recycling.

The scope of the soft strip will depend on the project and the design of the buildings involved; however, there are some common factors such as removal of carpets, fixtures and fittings. All works to be undertaken are programmed and form part of the main project programme.

A current typical sequence of activities for a soft strip might be as follows:

- Survey works and investigations that include the location and isolation of all existing redundant services. This is a complicated process and must be managed with a strict adherence to a pre-agreed strategy of location, isolation, proving, certifying and marking up.
- An intrusive survey for hazardous materials, including asbestos, and checks on registers are also essential to protect the health and safety of the operatives involved and other parties.
- Temporary lighting, power and emergency signage are installed for use during the soft-strip works.
- A first pass is undertaken to remove all carpets, curtains, excess furniture and accumulated rubbish, as well as doors and architraves and other fixtures and fittings. This first pass starts from the top of the building, working down, in order to allow other demolition activities to follow on in quick succession.
- Removal and safe disposal of hazardous materials, such as asbestos, usually undertaken by specialist contractors.
- During the first pass, the location, marking up and isolating of existing services is commenced, if not already undertaken. This may include protection and, if necessary, venting of rainwater pipes. Diversion of services may be required.
- Once services have been isolated, a second pass is made to remove all redundant services. This part of the soft-strip package may also include removal of non-structural elements such as partition walls. If any materials containing asbestos have been identified they are removed as a separate controlled process.
- Works affecting party walls and hoardings.

Current practice is to complete the demolition as quickly and efficiently as possible. Materials and components of significant value are generally removed carefully for reuse or recycling. Saving low-value materials for reuse or recycling is usually a secondary concern.

A key factor influencing the value of the materials and components is the ease with which they can be removed from the building. Where more effort is required to remove or separate them from other materials, then they are more likely to be left in situ and demolished during the physical removal of the structures. The value obtained by reusing or recycling materials may be reflected in a reduced price for undertaking the soft strip and other demolition works.

Careful deconstruction at the end of the building's useful life is likely to require a longer and more detailed first pass, as materials and components need to be removed from the building more carefully. Good design to facilitate deconstruction will make removal easier to achieve (and therefore faster), resulting in more materials being reused and recycled, thus reducing environmental impact and overall cost.

The materials streams of a soft strip

The waste streams that are identified will depend on the type of building that is to be stripped, for example a pharmaceutical plant compared to an office building. However, in broad terms, the items and materials removed in a soft strip are generally the lightweight items, including:

- building services plant and equipment;
- architectural features, including fireplaces, banisters, panelling, and so on;
- rubbish (both domestic and industrial);
- remaining furniture and equipment;
- kitchens, sanitary ware, and so on;
- lighting (bulbs, tubes, reflectors);
- light partition walls and suspended ceilings;
- doors, windows and frames;
- pipes and electrical installations;
- floors (carpets, tiles, wooden boards);
- roof materials.

The stripping methods used for the soft strip of the project will depend crucially on how the various elements are fixed – whether nailed, screwed, clipped, built in, built around, and so on. It also depends upon the types of material to be reclaimed for reuse or recycling or disposed of, and the possible disposal routes of the waste arising.

With this in mind, the contractor undertaking demolition should view the process in a holistic manner at the planning stage so that methods of work are complementary to all risks and hazards regarding the materials present, as well as the need to optimize the management of the waste streams that are produced. For the UK, BS 5618:2000 gives more information on risks and hazards.

When the waste streams have been established the contractor undertaking demolition can calculate the amounts of waste to establish the cost of disposal by the various available means, including those demanded by law. After a full cost analysis, the contractor compares the alternative routes of disposal, namely reuse, recycling of some form, energy recovery or landfill, and selects the most suitable ones.

Demolition

After the soft strip of a building, the sequence of building demolition and the detailed execution of the various stages depends on the size and type of the building, but typically is as follows:

- roofing tiles;
- roof beams;
- timber;
- internal plasterboard;
- exterior cladding;
- floors;
- major structural elements;
- footings and foundations.

These stages generally yield six main types of materials:

1. fixtures and fittings, including cables, ducts, and so on;
2. roofing materials;
3. plasterboard;
4. timber;
5. structural steel;
6. concrete and bricks.

Appropriate demolition processes can be established after consideration of the following:

- type and height of structure and nature of its construction (e.g. post-tensioned concrete);
- all services linked to the structure and within the structure's immediate location;
- health and safety and environmental risks associated with materials present in the structure;
- health and safety and environmental risks associated with demolition techniques;
- the local environment: local residents, environmentally sensitive areas, possible pollution targets;
- constraints on the use of plant and temporary works, including access to and egress from the site and pedestrian routes;
- nature and value of marketable components and materials;
- quantities of materials produced by the demolition process;
- management of materials streams on site;
- routes of disposal for waste streams;
- proximity of site to materials recycling facilities and waste-transfer stations;
- for partial demolition, any additional requirements with respect to other site occupants.

The efficiency and economic performance of the demolition process will be maximized by completion of a cost–benefit analysis addressing:

- legal waste disposal commitments (e.g. disposal of hazardous wastes);
- alternative opportunities for reuse of components and recycling of materials;
- environmental best practice;
- all alternatives to disposal as landfill.

The materials streams of demolition

Recycling is very much easier to do if all the different materials can be separated using multiple bins on site – glass, plasterboard, timber, paper, rubble, scrap, and so on. These bins all require space on site and good access for vehicles; this is not always possible on a construction site, in which case the materials may be removed to a waste-transfer station where they are separated for recycling.

The materials streams arising from a typical structure include the following:

- iron and steel, especially structural steel;
- non-ferrous metals;
- glass;
- roof covering;
- bricks and slates (structure or façade);
- cladding materials, such as masonry, concrete, glass-reinforced plastic, marble, and so on;
- masonry concrete (rubble);
- polluted concrete (chimney elements and so on);
- mechanical and electrical services and localized connections;
- pavement materials, including asphalt;
- soil, foundations and fill;
- non-combustibles (ceramic pipes, plaster, pumping, porcelain and insulating materials such as glass fibre);
- combustibles (timber, cardboard, polymers and paper).

Material recycling facilities/waste transfer stations

For many city-centre or large-town demolition and construction sites the shortage of space means that it is often difficult to segregate waste on site. This results in the use of skips or bins that will be filled with a mixture of waste materials and products. In order to capitalize on the potential reuse or recycling of this material, waste management companies have developed two types of facility for processing these materials off site.

Materials Recovery (or Recycling) Facilities

Materials Recovery (or Recycling) Facilities (MRF) are sites where source-segregated, dry, recyclable materials are mechanically or manually sorted to market specifications for further processing into secondary materials. MRFs specialize in dealing with waste whose composition is relatively well known, for example commercial or office materials such as paper and cardboard which are more easily matched to markets for secondary materials.

Mixed waste sorting facilities and waste-transfer stations

Mixed waste sorting facilities (MWSFs) and waste-transfer stations are also known as 'dirty' MRFs or dirty waste-transfer stations and are designed to process mixed waste that is unmanaged and may contain large volumes of 'wet' wastes such as soils and other organic materials. MWSFs are best suited to handling materials from construction sites.

Historically waste management facilities have been low-tech, labour-intensive operations. The mixed waste enters the site, passing through a weighbridge to calculate the weight of the waste (and therefore the charges applicable). The unsorted material is then deposited to a holding area where it is segregated into the various material types.

In such low-tech facilities the segregation of materials is achieved mainly using manual labour. Not all materials that are sorted can be recycled easily and operatives may be remunerated according to the volume of materials they segregate and sort for recycling.

In recent years a number of drivers are beginning to encourage mechanization and automation of the separation processes machinery – the costs of materials disposal, the introduction of increasingly stringent waste management legislation and the setting of national and local recycling targets. This is occurring despite the fact that the various markets for segregated materials still fluctuate considerably and very quickly. The equipment used in such modern, high-tech facilities may include:

- bag splitters to separate waste from containers;
- physical segregation using grabs or scoops;
- inclined conveyors to transport the materials through the facility;
- air separation to remove lightweight fractions;
- screening/trommel to remove fine materials;
- magnets to remove the ferrous fraction;
- paper screen/shredding equipment;
- balers for cardboard waste.

As more sophisticated and automated methods are introduced, so the number and range of materials that can be easily segregated are likely to increase in the future. This will improve the economic viability of reuse and recycling. Even in high-tech facilities, some hand-sorting may be required for lightweight materials such as plastics. Often the relatively small volumes of plastics received in unsegregated materials streams do not justify the high capital cost of installing automated separators.

Once sorted the materials streams may be:

- processed further at the facility, for example timber is often shredded and recycled as mulch or compost for landscaping, chipboard manufacture;
- given away, for example soil fractions passed through a trommel are stockpiled and supplied free of charge to anyone willing to take the material;
- sold on to other recycling organizations for further processing elsewhere, for example scrap metal is separated by type of metal and sold on to scrap dealers for smelting into new metal products;
- incinerated or sent to landfill, if not suitable for reuse or recycling.

B Assessing Environmental Impact

There are two main methods for assessing the environmental impacts of alternative building designs to compare the benefits to be gained from reuse and recycling:

- calculation from first principles;
- whole-building impact assessments, for example BREEAM or LEED.

Calculation of environmental impact from first principles

Life-cycle analysis

Life-cycle analysis (LCA) is a process that evaluates the environmental burdens associated with a product, process or activity. This involves identifying and quantifying energy and materials used and wastes released to the environment. LCA was developed in response to issues raised about environmental protection. Formerly the response to environmental concerns was to develop 'end of pipe' technologies that would aim to clean up the pollution produced at the end of a manufacturing process. It soon became clear that this would not be sufficient in the long term. Long-term pollution effects could only be addressed by finding out which parts of a process harm the environment and which do not. Life-cycle analysis has also led manufacturers and service providers to look at what happens to their goods once they have left their control, changing the emphasis from 'cradle to gate' to 'cradle to grave'. LAC has evolved in order to provide a reliable and standardized way of measuring environmental impacts (te Dorsthorst et al, 2000; Thormark, 2000; Berge, 2001).

Environmental profiles of materials

Different research organizations in different countries have devised their own methodologies for presenting environmental information to help reduce the confusion of claims and counterclaims about the performance of building materials. They are a useful way of providing reliable and independent environmental information about building materials and components in a standardized way – identifying and assessing the environmental effects of building materials over their entire life cycle, through their extraction, processing, construction, use and maintenance, and their eventual demolition and disposal.

In the UK the Building Research Establishment (BRE) has developed environmental profiles to enable designers to demand reliable and comparable environmental information about competing building materials, and give suppliers the opportunity to present credible environmental information about their products. Comparing different products on the basis of individual claims is difficult when there is no way of knowing if the assessment methods used in each case have considered the same factors. For example:

- Is the information based on typical UK practice?
- Does the information make predictions about recycling?
- Is transport included (and return journeys)?

Differences between such details can make comparisons of information from different sources meaningless. The environmental impact of materials is both varied and complex. The environmental profiles of materials can be assessed in two ways. It can be presented as a raw 'inventory' of data relating to inputs and outputs:

- material use;
- water use;
- emissions to air;

- emissions to water;
- embodied energy;
- emissions to land.

Alternatively the environmental impacts caused by the inputs and outputs can be considered:

- climate change;
- acid deposition;
- water pollution – eutrophication;
- water pollution – ecotoxicity;
- ozone depletion;
- minerals extraction;
- fossil-fuel depletion;
- water extraction;
- air pollution – human toxicity;
- air pollution – low-level ozone creation;
- waste disposal;
- transport pollution and congestion.

Ecopoints

The problem with the environmental impacts assessed in an environmental profile is that they are measured in different units and cannot be added together to calculate an aggregate impact. The environmental impacts of construction encompass a wide range of issues, including climate change, mineral extraction, ozone depletion and waste generation. Assessing such different issues in combination requires subjective judgements about their relative importance. For example, does a product with a high global warming impact that does not pollute water resources have less overall environmental impact than a product that has a low global warming impact but produces significant water pollution? To enable such assessments, the BRE in the UK has developed a common unit of impact called the 'ecopoint' (Figure A1.1).

Expert panels from across the industry's stakeholder groups were asked to judge the importance of different environmental impacts and relative weightings were assigned. The resulting weightings for the environmental issues measured by the BRE have been used to weight the normalized environmental impacts to provide the ecopoints score.

One hundred UK ecopoints is defined as the annual environmental impact caused by a typical UK citizen. Research organizations in other countries have developed their own version of the ecopoint. One UK ecopoint is equivalent to:

- using 320kWh of electricity;
- using 83m^3 of water (enough to fill 1000 baths);
- travelling 65 miles by articulated truck;
- landfilling 1.3 tonnes of waste;
- manufacturing 0.75 tonnes of brick (250 bricks);
- transporting 540 tonne-kilometres by sea freight;
- using 1.38 tonnes of mineral extraction;
- undertaking 300 miles of urban driving in a new petrol-driven car.

Whole-building impact assessments

BREEAM-for-Offices 2005 and EcoHomes

BREEAM-for-Offices and EcoHomes (for residential buildings) have both been devised by the Building Research Establishment for use in the UK. The assessments involve around 70 criteria leading to a percentage score and a rating (excellent, very good, good or pass). The criteria relating to the reuse of building elements and the use of recycled materials are very similar for each assessment method (Table A1.1).

LEED (US Green Building Council)

The Leadership in Energy and Environmental Design (LEED™) Green Building Rating System represents the US Green Building Council's effort to provide a national standard for what constitutes a 'green building'. Through its use as a design guideline and third-party certification tool, it aims to improve occupant well-being, environmental performance and economic returns of buildings using established and innovative practices, standards and technologies. A total of 69 points are available and banded to give a rating – platinum, gold or silver (Table A1.2).

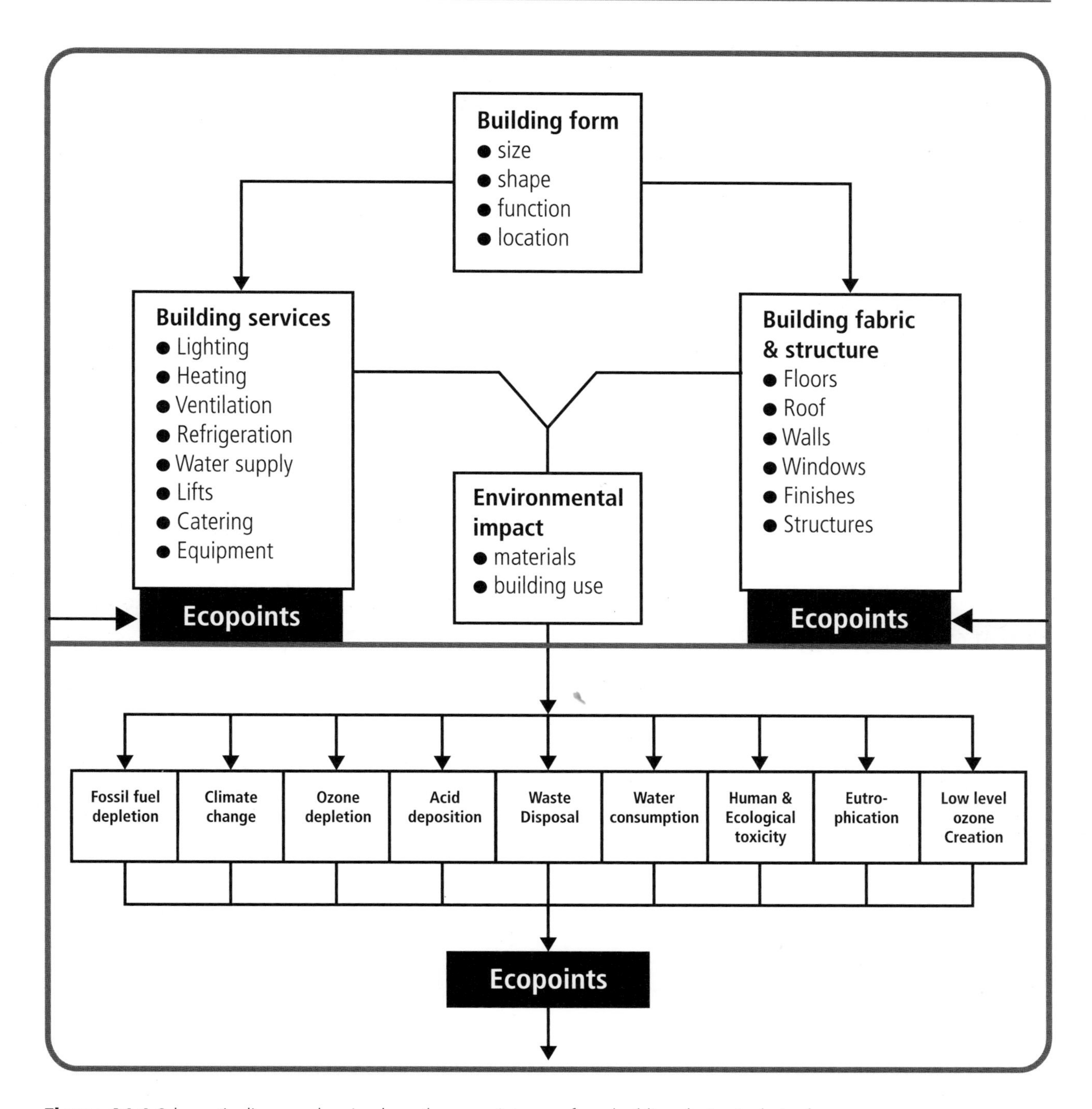

Figure A1.1 Schematic diagram showing how the ecopoint score for a building design is derived

Source: Building Research Establishment

Table A1.1 Credits in BREEAM-for-Offices (2005) and EcoHomes relating to reuse/recycling

Credit category	Criteria	Credits
Mat 1.1	Reuse of a building façade One credit is awarded where at least 50% of the total façade (by area) is reused and at least 80% of the reused façade (by mass) comprises in situ reused material	1
Mat 1.2	Reuse of structure One credit is awarded where a design reuses at least 80% of an existing primary structure by gross building volume. Where a project is part refurbishment and part new build, the volume of the reused structure must comprise at least 50% of the final structure's volume	1
Mat 1.6	Recycled aggregates One credit is awarded where significant use of crushed aggregate, crushed masonry or alternative aggregates (manufactured from recycled materials) are specified to deliver positive aspects of the design (such as the building structure, ground slabs, roads etc.)	1
Mat 1.7	Sustainable timber Up to two credits are available where timber and wood products used in structural and non-structural elements are responsibly sourced, OR utilize reused or recycled timber. Recycled/reused timber can be introduced as part of the proportion for certified timber. In a refurbished building, existing timber is treated as reused. (Any recycled/reused timber need not be from a 'sustainable' source)	Up to 2

GBTool (Green Building Challenge)

GBTool is the method developed for use in the international Green Building Challenge (GBC) to assess the potential energy and environmental performance of building projects A feature of GBTool that sets it apart from other assessment systems is that the method is designed from the outset to allow users to reflect the very different priorities, technologies, building traditions and even cultural values that exist in various regions and countries (Table A1.3). The GBC process is managed by the International Initiative for a Sustainable Built Environment (IISBE) (see www.iisbe.org).

Table A1.2 Credits in LEED version 2.1 relating to reuse/recycling

Credit category	Criteria	Credits
	Building reuse	
1.1	Maintain at least 75% of existing building structure and shell (exterior skin and framing, excluding window assemblies)	1
1.2	Maintain an additional 25% (100% total) of existing building structure and shell (exterior skin and framing, excluding window assemblies and non-structural roofing material)	1
1.3	Maintain 100% of existing building structure and shell (exterior skin and framing, excluding window assemblies and non-structural roofing material) AND at least 50% of non-shell areas (interior walls, doors, floor coverings and ceiling systems)	1
	Construction waste management	
2.1	Develop and implement a waste management plan, quantifying material diversion goals. Recycle and/or salvage at least 50% of construction, demolition and land-clearing waste. Calculations can be done by weight or volume, but must be consistent throughout	1
2.2	Develop and implement a waste management plan, quantifying material diversion goals. Recycle and/or salvage an additional 25% (75% total) of construction, demolition and land-clearing waste. Calculations can be done by weight or volume, but must be consistent throughout	1
	Resource reuse	
3.1	Use salvaged, refurbished or reused materials, products and furnishings for at least 5% of building materials	1
3.2	Use salvaged, refurbished or reused materials, products and furnishings for at least 10% of building materials	1
	Recycled content	
4.1	Use materials with recycled content such that post-consumer recycled content constitutes at least 5% of the total value of the materials in the project OR combined post-consumer and half post-industrial recycled content constitutes at least 10%; The value of the recycled content portion of a material or furnishing shall be determined by dividing the weight of recycled content in the item by the total weight of all material in the item, then multiplying the resulting percentage by the total value of the item; Mechanical and electrical components shall not be included in this calculation; Recycled content materials shall be defined in accordance with the Federal Trade Commission document, *Guides for the Use of Environmental Marketing Claims*	1
4.2	As 4.1, but 20% instead of 10%	1

Table A1.3 Credits in GBTool (Version 1.82, 2002) relating to reuse/recycling

Credit	Criteria	Performance measure	Score
R4	Reuse of existing structure or on-site materials	Assesses the extent to which on-site material resources have been reused. Non-applicable if there are no existing structures on site that are of sufficient size or quality for the intended function	
R4.1	Retention of an existing structure on the site; This criterion assesses the extent to which any existing on-site building is integrated into the new building	The proportion of the structure of an existing building on the site that is retained as part of the new building: percentage of gross floor area (0–90%)	–2 to +5
R4.2	Off-site reuse or recycling of steel from existing structure on the site; This criterion assesses the amount of salvageable steel in an existing structure (if applicable) that has been reused on site or recycled off site (only applicable in the case of an existing steel-framed building on site)	The weight of steel taken from the part of the existing structure not reused in the new building: percentage of total weight (39–78%); This assessment excludes any materials accounted for in Criterion 4.1	–2 to +5
R4.3	Off-site reuse of materials from existing structure on the site; This criterion assesses the extent to which salvageable materials in an existing structure (if applicable) have been reused on site or recycled off site.	The proportion of materials and components salvaged from existing structure(s) on the site that are reused in the new or renovated structure: percentage of total weight (0–100%); This assessment excludes any materials accounted for in Criterion 4.1	–2 to +5
R5	Amount and quality of off-site materials used	Covers the environmental attributes of materials in the building originating off site	
R5.1	Use of salvaged materials from off-site sources; This criterion assesses the amount of salvaged materials and components originating from the demolition or refurbishment of buildings off site.	The proportion of the materials and components used in the case study building that are salvaged from off-site sources: percentage of total weight of materials in new building (3–10%)	–2 to +5
R5.2	Recycled content of materials from off-site sources; This criterion covers the recycled content of the major architectural materials used in the case-study building	The proportion of architectural materials and components that have recycled content: percentage of total weight (3–10%); Different materials and components will have different recycled contents, and this may be either post-industrial or post-consumer in origin. The assessment is based on the aggregate of the material weight multiplied by their respective recycled contents	–2 to +5
R5.3	Use of wood products that are certified or equivalent; This criterion covers all wood products used in the building including framing, flooring, finishes and millwork	The proportion of materials of wood origin certified to conform to requirement for sustainable forestry practice guidelines: percentage of the weight of all wood-based components (15–33%) (salvaged timber is counted as certified timber)	–2 to +5
S5.2	Quality of parking area development; This criterion assesses the extent to which the design includes measures to minimize the adverse affects of on-grade [i.e. ground level] parking areas	Types of strategies (involving reused/recycled materials) rewarded include: using crushed stone or brick for lightly used pedestrian paths; using recycled asphalt and recycled concrete where impervious surfaces are required	–2 to +5

Note: Scores are weighted differently; each point is worth much less than 1 per cent of the final total.

C Glossary

deconstruction	a process of carefully taking apart components of a building, possibly with some damage, with the intention of either reusing some of the components after refurbishment or reconditioning, or recycling the materials. It may be undertaken during refurbishment, when adapting a building for new use, or at the end of its life
Delft Ladder	a sequential model of the life cycle of materials and products identifying different stages of use and degradation towards waste, developed at the University of Delft
demolition	a term for both the name of the industry and the process of intentional dismantling and reduction of a building, or part of a building, without necessarily preserving the integrity of its components or materials
design for deconstruction	the process of designing a building to facilitate its deconstruction or disassembly. The same idea is sometimes conveyed as 'design for disassembly', and both are often abbreviated as DfD
disassembly	a synonym for dismantling; an antonym of assembly
dismantling	a reversible process of taking apart components of an artefact, without damaging them, with the intention of reassembly of the entire artefact, for example for maintenance, reconditioning, remanufacture or re-erection in a new location
down-cycle	reuse a product, component or material for a purpose with lower performance requirements than it originally provided
energy recovery	incineration of waste to generate energy
post-consumer waste	material surplus to requirements following its use by consumers, for example plastic bottles, crushed concrete
post-industrial waste	material surplus to requirements generated during manufacturing processes, for example timber off-cuts or sawdust from a timber mill, pulverized fuel ash
primary material	a material whose production has involved extraction from natural reserves
reclaimed material	material extracted from the waste stream and reused without major processing
reclamation	the collection and separation of materials from the waste stream
recondition	the process of restoring a building element or piece of equipment to a condition that allows it to be reused
recovery	use of waste materials in order to prevent their disposal to landfill, usually by recycling, composting or energy recovery

recycle	collect and separate usable materials from waste and process them to produce marketable products
recycled content	the proportion of recycled materials used to make a product
RCBP	recycled-content building product; a product made using a proportion of recycled materials for use in building construction
recycled material	material that is processed to produce a derivative product
refurbishment	any alteration that is intended to improve a building, ranging from redecoration to rearrangement of partition walls, installation of new building services or lifts, roof or façade replacement, to moving load-bearing columns or walls; also the improvement of a second-hand product making it suitable for reuse
remanufacture	bringing an artefact back into use by means of deconstruction or disassembly followed by processes similar to those used for its original manufacture and assembly; for example a used steel beam being cut to length and provided with means for making a new connection detail to enable it to be used for a second time
reuse	putting objects back into use, either for their original purpose or a different purpose without major prior reprocessing to change their physical characteristics, in order that they do not enter the waste stream. While it does not include reprocessing, it might involve some reconditioning
salvaged product	a product removed from a building during deconstruction or demolition to be reused after little or no refurbishment or remanufacture
secondary material	material that has previously been used for a primary purpose
soft strip	the initial stage of the demolition process during which high-value and easily removable items are taken out, prior to demolition of the building structure
trommel	a rotary, cylindrical sieve for separating waste materials by size
waste	the everyday meaning of the word 'waste' is something between material that is discarded after use, or simply unwanted material. Legal definitions become very detailed but add little of relevance to its use in this guide
waste arisings	total quantities of waste generated from any process or activity
waste hierarchy	ranking of alternative means of waste management, in sequence of increasing severity of environmental impact

References

Addis, W. and Talbot, R. (2001) *Sustainable Construction Procurement: A Guide to Delivering Environmentally-responsible Projects*. Report C571. London: CIRIA

Addis, W. and Schouten, J. (2004) *Design for Deconstruction: Principles of Design to Facilitate Reuse and Recycling*. Report C607. London: CIRIA

Anink, D., Boonstra, C. and Mak, J. (1996) *Handbook of Sustainable Building: An Environmental Preference Method for Selection of Materials for Use in Construction and Refurbishment*. London: James & James

Anon (2001) 'Using recycled aggregates: Wessex Water Operations Centre, Bath', *The Structural Engineer*, vol 79, no 12, pp16–18

Beckmann, P. (1995) *Structural Aspects of Building Conservation*. Maidenhead: McGraw Hill

Berge, B. (2001) *The Ecology of Building Materials*. Oxford: Architectural Press

Biffa Waste Services (2002) *Future Perfect*. High Wycombe: Biffa

Bitsch Olsen, E. (1993) 'The recycled house in Odense', 3rd International RILEM Symposium on Demolition and Reuse of Concrete and Masonry, Odense, Denmark

Bowman, M. D. and Betancourt, M. (1991) 'Reuse of A325 and A490 high-strength bolts', *Engineering Journal*, vol 28, no 3, pp17–26 (Reprinted by American Institute of Steel Construction (AISC), 2002)

BRE (1991) *Structural Appraisal of Existing Buildings for Change of Use*. BRE Digest 366. Garston: Building Research Establishment

BRE (1997) *Plastics Recycling in the Construction Industry*. Info Sheet No12/1997. Garston: Building Research Establishment

BRE (1998) *Recycled Aggregates*. Digest 443. Garston: Building Research Establishment

BRE (2003) *Concrete in Aggressive Ground. Part 1: Assessing the Aggressive Chemical Environment*. Special Digest 1. Garston: Building Research Establishment

BS 6543:1985 (1985) *Guide to the Use of Industrial By-products and Waste Materials in Building and Civil Engineering*. London: HMSO

BS 7671:2001 (2001) IEE *Wiring Regulations*. 16th Edition. London: HMSO

Bussell, M. N. (1997) *Appraisal of Existing Iron and Steel Structures*. SCI Publication 138. Ascot: Steel Construction Institute

Bussell, M. N. and Robinson, M. J. (1998) 'Investigation, appraisal and reuse of a cast-iron structural frame', *The Structural Engineer*, vol 76, no 3, pp37–42

Bussell, M. N., Lazarus, D. and Ross, P. (2003) *The Retention of Masonry Façades: Best Practice Guidance*. Report C579. London: CIRIA

Chapman, T., Marsh, B. and Foster, A. (2001) 'Foundations for the future, proceedings of Institution of Civil Engineers', Paper 12340, *Civil Engineering*, vol 144, pp36–41

Chow, F. C., Chapman, T. J. P. and St John, H. D. (2002) 'Reuse of existing foundations: Planning for the future', *Proceedings of the 2nd International Conference on Soil Structure Interaction in Urban Civil Engineering*. Zurich

CIBSE (1999) *Guide A: Environmental Design*. London: Chartered Institute of Building Services Engineers

CIRIA (1994) *Structural Renovation of Traditional Buildings*, Report 111 (revised edition). London: Construction Industry Research and Information Association

CIRIA (2006) *Reuse of Foundations for Urban Sites: Best Practice Handbook* (6653). London: CIRIA

Coles, B., Henley, R. and Hughes, R. (2001) 'The reuse of pile locations at Governor's House development site, City of London', *Proceedings of Conference on Preserving Archaeological Remains in Situ*. Museum of London

Archaeological Service and University of Bradford
Collins, R. J. (1994) *The Use of Recycled Aggregates in Concrete*. BRE IP5/94, Garston: Building Research Establishment
Collins, R. J. and Sherwood, P. T. (1995) *Use of Waste and Recycled Materials as Aggregates: Standards and Specifications*. Report for the Department of the Environment. London: HMSO
Collins, R. J., Harris, D. J. and Sparkes, W. (1998) *Blocks with Recycled Aggregate: Beam and Block Floors*. BRE IP14/98. Garston: Building Research Establishment
Courtney, M. and Matthews, R. (1988) 'Tobacco Dock', *Arup Journal*, vol 23, no 3, pp6–11
Coventry, S. and Guthrie, P. (1998) *Waste Minimisation and Recycling in Construction: Design Manual*. Special Publication 134. London: CIRIA
Coventry, S., Woolveridge C. and Hillier, S. (1999) *The Reclaimed and Recycled Construction Materials Handbook*. Report C513. London: CIRIA
Dean, A. and O. Hursley, T. (2002) *Rural Studio: Samuel Mockbee and an Architecture of Decency*, Princeton: Princeton Architectural Press
de Vries, P. (1993) 'Concrete recycled: Crushed concrete as aggregate', *Concrete*, vol 27, no 3, pp9–13
Durmisevic, E. and Noort, N. (2003) 'Re-use potential of steel in building construction', CIB Publication 287, *Deconstruction and Materials Reuse*. Proceedings of the 11th Rinker International Conference, 7–10 May, Gainesville, Florida, University of Florida
Eklund, M., Dahlgren, S., Dagersten, A. and Sundbaum, G. (2003) 'The conditions and constraints for using reused materials in building projects', CIB Publication 287, *Deconstruction and Materials Reuse*. Proceedings of the 11th Rinker International Conference, May 7–10, Gainesville, Florida, University of Florida
EPA (2000) *Comprehensive Procurement Guidelines*. Washington DC: United States Environmental Protection Agency
Friedman, D. (1995) *Historical Building Construction: Design, Materials and Technology*. (US Practice) New York: W. W. Norton
Gorgolewski M. (2000) 'The Recycled Building Project', *Conference Proceedings of Sustainable Building* 2000, Maastrict, Holland
Hobbs, G. and Collins, R. J. (1997) *Demonstration of Reuse and Recycling*. BRE IP3/97. Garston: Building Research Establishment
HVCA (2000) A *Practical Guide to Ductwork Leakage Testing*. Report DW/143. London: Heating and Ventilating Contractors' Association
ICE (2004) *Demolition Protocol*. London: Institution of Civil Engineers
IStructE (1996) *Appraisal of Existing Structures*. Second Edition. London: Institution of Structural Engineers
IStructE (1999) *Building for a Sustainable Future: Construction without Depletion*. London: Institution of Structural Engineers
IWMB (2000) A *Technical Manual for Material Choices in Sustainable Construction*. State of California: Integrated Waste Management Board
Kay, T. (2000) *The Salvo Guide* 2000. Berwick-upon-Tweed: Salvo
Kernan, P. (2002) *Old to New: Design Guide – Salvaged Materials in New Construction*. Third Edition. Vancouver: Greater Vancouver Regional District Policy and Planning Department
Kibert, C. J. (1993) 'Construction materials from recycled polymers', *Proceedings of the Institution of Civil Engineers*, vol 99, no 4, pp455–464
Kristinsson, J., Hendricks, C. F., Kowalczyk, T. and te Dorsthorst, B. J. H. (2001) 'Reuse of secondary elements: Utopia or reality', CIB *World Building Congress*, April 2001, Wellington, New Zealand, Paper No 230
Lazarus, N. (2002) *Beddington Zero (Fossil) Energy Development: Construction Materials Report: Toolkit for Carbon Neutral Developments*, Part 1. Wallington, Surrey: Bioregional Development Group (part available at www.bioregional.com and www.bedzed.org.uk)
MacDonald, S. (ed) (2003) *Concrete Building Pathology*. London: Blackwell Publishing
McGrath, C., Fletcher, S. L. and Bowes, H. M. (2000) 'UK Deconstruction Report', CIB *Publication* 252: *Overview of*

Deconstruction in Selected Countries, *Proceedings of* CIB *Task Group* 39, Florida

Mitchell, J. M., Courtney, M. and Grose, W. J. (1999) 'Timber piles at Tobacco Dock, London', *Proceedings of* ISSMFE *Conférence Colloque International Fondations Profondes*, Paris

NBS (1997) *Greening the National Building Specification: The Greening Report*. London: National Building Specification (available at www.ecde.co.uk)

NES (2001) *A Guide to Sustainable Engineering Specification*. Windsor: Barbour Index (available at www.ecde.co.uk)

NGS (2004) *Sustainability Checklist*. National Green Specification/Brian Murphy (available at www.greenspec.co.uk)

Olnhoff, V. and Martin, A. (2003) *Recycling Building Services*. BG 16/2003. Bracknell: Building Services Research and Information Association

Price, S. and Rogers, H. (2005) 'Stone Cantilever Staircases', *The Structural Engineer*, vol 83, no 2, pp29–36

Rayner, J. (ed) (2002) *Materials Recycling Handbook*. Croydon: EMAP

Robinson, M. and Marsland, A. (1996) 'Canalside West, Huddersfield', *Arup Journal*, vol 3, pp12–14

Ross, P. (2002) *Appraisal and Repair of Timber Structures*. London: Thomas Telford

Roth, L. and Eklund, M. (2001) 'Environmental analysis of reuse of cast-in-situ concrete in the building sector', in Shanalbleh, A. and Chang, W. P. (eds) *Towards Sustainability in the Built Environment*. Brisbane, Australia: Faculty of the Built Environment and Engineering, Queensland University of Technology, pp234–243

SCI (2000) *Concept Study and Economic Assessment of a Recycled Building Using a Primary Steel Frame*. Ascot: Steel Construction Institute

St John, H. D., Chow, F. C. and Freeman, T. J. (2000) 'Follow these footprints', *Ground Engineering*, December, vol 33, pp24–25

Swailes, T. and Marsh, J. (1998) *Structural Appraisal of Iron-Framed Textile Mills*. London: Thomas Telford

te Dorsthorst, B. J. H., Kowalczyk, T., Hendriks, C. F. and Kristinsson, J. (2000) 'From grave to cradle: reincarnation of building materials', *Conference Proceedings of Sustainable Building* 2000, Maastrict, Holland

te Dorsthorst, B. J. H. and Kowalczyk, T. (2002) 'Reuse versus demolition: A case study in the Netherlands', *Conference Proceedings of Sustainable Building* 2002, Oslo, Norway

Thormark, C. (2000) 'Environmental analysis of a building with reused building materials', *International Journal of Low Energy and Sustainable Buildings*, vol 1, www.byv.kth.se/avd/byte/leas

Thornton, A. (1994) 'Relocation of Museum Hotel', *Transactions of the Institution of Professional Engineers New Zealand*, vol 21, (1/CE), November, pp1–6

Wong, D. J. and Perkins, C. (2002) 'The integrated Arup Campus', *New Steel Construction*, vol 10, no 1, January/February, pp21–23

Woolley, T., Kimmins, S., Harrison, R. and Harrison, P. (1997) *Green Building Handbook: A Guide to Building Products and Their Impact on the Environment, Volume* 1. London: Spon Press

Woolley, T. and Kimmins, S. (2000) *Green Building Handbook: A Guide to Building Products and Their Impact on the Environment, Volume* 2. London: Spon Press

Yeomans, D. (2003) *The Repair of Historic Timber Structures*. London: Thomas Telford

Organizations and Websites

The quality and speed of internet search engines are now so high that it is hardly necessary to give references to websites. Nevertheless, a number of organizations and their websites are listed below to provide the reader with an introduction to a sample of the many and various types of organization that are engaged in the world of reclamation, reuse and recycling.

Suppliers of reclaimed/salvaged goods and materials

Most countries have many architectural salvage firms and traders in reclaimed goods and materials. In the UK, one way of locating such firms is through the Salvo website that provides useful search options:

- www.salvo.co.uk

A directory of suppliers is provided on Kingston University website:

- www.recyclingbydesign.org.uk

In Holland, the firm Bouwcarrousel provides a reclamation service that ranges from salvaging goods and buildings being demolished, to selling goods refurbished to the desired level:

- www.bouwcarrousel.nl/

Office and domestic furniture is a popular area of trade, not least because the products are easily movable and available in large quantities. Some examples are:

- www.reuze.co.uk/frn_directory.shtml (The Furniture Recycling Network)
- www.frn.org.uk/ (The Furniture Reuse Network)
- www.century-office-equipment.co.uk/recon.htm (Reconditioned office furniture)

Materials exchange

The internet provides a convenient means of trying to match supply and demand in reclaimed building materials. Two UK examples are:

- www.salvomie.co.uk/ (Materials Information Exchange)
- www.ciria.org/recycling/ (Construction recycling sites)

Suppliers of recycled-content building products

Various local governments now promote recycling through online databases of RCBPs, for example:

- www.ciwmb.ca.gov/RecycleStore (State of California, USA)
- www.oaklandpw.com/oakrecycles/construction/products.htm (City of Oakland, California, USA)
- www.state.nj.us/dep/dshw/recyclenj/building.htm (New Jersey, USA)

Various online databases provide links to suppliers of recycled-content products:

- www.GreenBuildingStore.co.uk (The Green Building Store)
- www.recycledproducts.org.uk (Organised by Waste and Resources Action Programme, WRAP)
- www.ecoconstruction.org/ EcoConstruction Database

In the UK, the Waste and Resource Action Programme (www.wrap.org.uk) provides guidance on using some recycled materials, as raw materials or in products:

- www.aggregain.org.uk (recycled aggregate for concrete)
- www.recyclewood.org.uk (recycled timber)

Materials and product specification

In the UK, the National Green Specification provides a large range of information to help designers and specifiers of reclaimed and recycled materials:

- www.greenspec.co.uk/

In the US, a number of local authorities already provide guidance on using reclaimed good and recycled materials. For example, in Seattle, the Department of Planning and Development has published a guide to *Sustainable Building and Reuse of Building Materials*, (Client assistance memo no 366, 2001, Seattle, US). This can be downloaded from www.ci.seattle.wa.us/dclu/Publications/cam/cam336.pdf.

A list of guides to specification for recycling is available at the 'Recycling-by-design' website:

- www.recyclingbydesign.org.uk/site/content/specifying.asp

Trade and professional organizations promoting reclamation, reuse and recycling

Guidance on particular materials is often available from relevant trade associations:

- www.recycle-it.org (Timber Recycling Information Centre)
- www.recoup.org (Recoup (plastics recycling))
- www.icer.org.uk (Industry Council for Electronic Equipment Recycling)
- www.bsria.org.uk (Building Services Research and Information Association)
- www.steel-sci.org/ (Steel Construction Institute)

Research organizations

Reclamation, reuse and recycling are subjects that have attracted the attention of various research organisations. The following is a small selection.

The Conseil International du Bâtiment (International Council for Building) (www.cibworld.nl) has a Technical Group (TG39) focusing on design for deconstruction that has many useful case studies:

- www.cce.ufl.edu/affiliations/cib/index.html

In the Netherlands a research project on industrial, flexible and demountable building (IFD – Industrieel Flexibel en Demontabel Bouwen) provides useful case studies on design for deconstruction:

- www.sev.nl/ifd

Many national building research institutes are undertaking research on reclamation, reuse and recycling, for example:

- Building Research Establishment (UK) (www.bre.co.uk);
- Belgian Building Research Institute (BBRI) (http://www.recyhouse.be/index.cfm?lang=en).

Kingston University in the UK has undertaken several projects on reuse and recycling in construction and other industries. Among various outcomes are included:

- guidance for designers (http://www.recyclingbydesign.org.uk/site/content/home.asp);
- inspiring product manufacture with recycling (www.inspirerecycle.org).

A series of international conferences called 'Sustainable Building', run every two to three years since 1997, have featured many papers on reclamation, reuse and recycling.

Index